GREAT DISASTERS

EARTHQUAKES

Nancy Harris, *Book Editor*

Daniel Leone, *President*
Bonnie Szumski, *Publisher*
Scott Barbour, *Managing Editor*

GREENHAVEN
PRESS ®

THOMSON

GALE

San Diego • Detroit • New York • San Francisco • Cleveland
New Haven, Conn. • Waterville, Maine • London • Munich

THOMSON

*

GALE

LIBRARY OF CONGRESS CATALOGING-IN-PUBLICATION DATA
Earthquakes / Nancy Harris, book editor.
p. cm. — (Great disasters)
Includes bibliographical references and index.
ISBN 0-7377-1648-7 (pbk. : alk. paper) — ISBN 0-7377-1647-9 (lib. : alk. paper)
1. Earthquakes. I. Harris, Nancy. II. Great disasters (Greenhaven Press)
QE534.3 .E37 2003
363.34'95—dc21   2002034655

Printed in the United States of America

CONTENTS

Chapter 1: The Science and Study of Earthquakes

Chapter 2: Earthquake Disasters

Chapter 3: Averting Disaster

tions and the importance of a continued search for survivors for at least two weeks after a catastrophe.

After the Loma Prieta earthquake in 1989, the State of California redoubled its earthquake mitigation and management efforts, relying heavily on seismic hazard maps created by a sophisticated geographic information system (GIS).

The public can answer questions on the Internet about their experience immediately following an earthquake disaster. Using this information, experts can quickly create maps to assess the earthquake's severity and to assist in emergency operations.

FOREWORD

Humans have an ambivalent relationship with their home planet, nurtured on the one hand by Earth's bounty but devastated on the other hand by its catastrophic natural disasters. While these events are the results of the natural processes of Earth, their consequences for humans frequently include the disastrous destruction of lives and property. For example, when the volcanic island of Krakatau exploded in 1883, the eruption generated vast seismic sea waves called tsunamis that killed about thirty-six thousand people in Indonesia. In a single twenty-four-hour period in the United States in 1974, at least 148 tornadoes carved paths of death and destruction across thirteen states. In 1976, an earthquake completely destroyed the industrial city of Tangshan, China, killing more than 250,000 residents.

Some natural disasters have gone beyond relatively localized destruction to completely alter the course of human history. Archaeological evidence suggests that one of the greatest natural disasters in world history happened in A.D. 535, when an Indonesian "supervolcano" exploded near the same site where Krakatau arose later. The dust and debris from this gigantic eruption blocked the light and heat of the sun for eighteen months, radically altering weather patterns around the world and causing crop failure in Asia and the Middle East. Rodent populations increased with the weather changes, causing an epidemic of bubonic plague that decimated entire populations in Africa and Europe. The most powerful volcanic eruption in recorded human history also happened in Indonesia. When the volcano Tambora erupted in 1815, it ejected an estimated 1.7 million tons of debris in an explosion that was heard more than a thousand miles away and that continued to rumble for three months. Atmospheric dust from the eruption blocked much of the sun's heat, producing what was called "the year without summer" and creating worldwide climatic havoc, starvation, and disease.

As these examples illustrate, natural disasters can have as much impact on human societies as the bloodiest wars and most chaotic political revolutions. Therefore, they are as worthy of study as the major events of world history. As with the study of social and political events, the exploration of natural disasters can illuminate the causes of these catastrophes and target the lessons learned about how to mitigate and prevent the loss of life when disaster strikes again. By examining these events and the forces behind them, the Greenhaven Press Great Disasters series is designed to help students better understand such cataclysmic events. Each anthology in the series focuses on a specific type of natural disaster or a particular disastrous event in history. An introductory essay provides a general overview of the subject of the anthology, placing natural disasters in historical and scientific context. The essays that follow, written by specialists in the field, researchers, journalists, witnesses, and scientists, explore the science and nature of natural disasters, describing particular disasters in detail and discussing related issues, such as predicting, averting, or managing disasters. To aid the reader in choosing appropriate material, each essay is preceded by a concise summary of its content and biographical information about its author.

In addition, each volume contains extensive material to help the student researcher. An annotated table of contents and a comprehensive index help readers quickly locate particular subjects of interest. To guide students in further research, each volume features an extensive bibliography including books, periodicals, and related Internet websites. Finally, appendixes provide glossaries of terms, tables of measurements, chronological charts of major disasters, and related materials. With its many useful features, the Greenhaven Press Great Disasters series offers students a fascinating and awe-inspiring look at the deadly power of Earth's natural forces and their catastrophic impact on humans.

One of the most destructive earthquakes in history, reported at 8.2 on the Richter scale, struck the city of Tangshan, China, at 3:45 A.M. on July 27, 1976. Official estimates list the death toll at 250,000, but it is believed that 650,000 to 750,000 people perished in the quake, an additional 500,000 were injured, and 800,000 were left homeless. The industrial city of Tangshan, with a population of 1 million, was almost completely demolished. The physical damage to the city has been compared to the destruction in Hiroshima, Japan, after it was devastated by a U.S. atomic bomb at the end of World War II.

An aftershock of magnitude 7.9 occurred in Tangshan several hours later. Each of the two shocks was roughly equivalent in magnitude to the earthquake that devastated San Francisco in 1906. The Tangshan disaster happened so close to the capital city of Beijing that officials ordered the citizens to move outdoors in case an earthquake struck there. An estimated 6 million people slept outdoors in temporary shelters for more than two weeks. China has had the greatest mortality rates due to earthquakes of any country. A much earlier quake in 1556 in the region of Shansi, China, killed more than eight hundred thousand people, the highest death toll from an earthquake ever recorded.

Although it did not cause as many fatalities as the Tangshan quake, the San Francisco earthquake of 1906 was one of the most infamous catastrophes in history. Estimated at 8.2 on the Richter scale, the quake created a surface rupture that extended almost three hundred miles. The tremors caused cooking stoves and kerosene lamps to topple over, resulting in fires that burned out of control and spread quickly through the ill-prepared city. The earthquake also broke water mains, leaving little water to fight the fires. An inferno blazed for three days and consumed almost five hundred blocks, defeating the army's efforts to contain it. Initial reports claimed that 315 people had perished in the San Francisco quake, but that number has since climbed to possibly 3,000.

These two examples demonstrate the scale of an earthquake's

destructive power. Earthquakes are the strongest and most destructive forces on earth. Major earthquakes can destroy entire
cities, killing thousands of people with massive jolts lasting less
than a minute. Earthquakes account for approximately 8,000
deaths and 26,000 injuries globally each year. According to the
United States Geological Survey (USGS), every year an average
of 18 major quakes and 49,000 minor ones strike somewhere in
the world. About 1,000 very minor quakes happen daily. Although earthquakes can strike anywhere, at any time, about 81
percent of the world's largest earthquakes occur around the rim
of the Pacific Ocean in an area called the circum-Pacific seismic belt.

A major earthquake is one that measures 7 to 7.9 on the
Richter scale. The Richter scale measures the strength, or magnitude, of a quake. However, many factors besides magnitude determine how much damage an earthquake will cause. Major
quakes that occur in remote areas are never reported, but others
that strike in or near heavily populated areas can cause property
damage in the hundreds of millions or even billions of dollars.

The Ring of Fire

One such area is Tokyo, Japan, where quakes have rocked the city in roughly seventy-five-year intervals, the last one having occurred in 1928. It is believed that the next major quake to strike Tokyo would cost $1 trillion and have a drastic worldwide economic impact.

An earthquake's duration is another factor affecting the level of damage. Some earthquakes last only a few seconds while others last one or two minutes. Longer quakes are generally more destructive. Much of the damage and death from an earthquake is due to collapsing buildings and flying debris, so the architecture in an area plays a vital role in earthquake damage as well. In general, wood-frame buildings, which can bend under earthquake vibrations, have a better chance of remaining upright. Brick and stone, by contrast, have a tendency to collapse and crumble, a situation that can be improved with better design and construction. Adobe is possibly the worst building material in the western United States, and was responsible for many fatalities in early earthquakes.

No matter what a building's construction material, the nature of the underlying ground makes a significant difference in whether a structure will remain standing or not. Solid rock is the safest foundation for damage protection. Less consolidated ground, when shaken, may liquefy. When this happens, groundwater rises to the surface, causing extensive damage as buildings tilt and collapse on the weakened ground. In addition, unbroken underground geological formations carry shock waves for long distances, maintaining the waves' initial intensity. For example, in 1812, the New Madrid Fault in Missouri ruptured and shook most of the eastern United States. In contrast, fractured earth tends to absorb shock waves and thus reduce the geographical extent of damage. For example, most of the effects of the San Andreas Fault are confined to California.

In the face of such destructive power, scientists have worked for centuries to try and predict where and when earthquakes will happen. Quake forecasting is still maturing as a science. Like weather forecasts, earthquake predictions indicate that a quake has a certain probability of occurring within a given time, not that one will definitely strike. Scientists can make these forecasts because earthquakes tend to occur in clusters that strike the same area in a particular period of time. The smaller foreshocks of an earthquake cluster occur prior to the main shock and are one of the clues scientists use to make predictions. Other warning signs include ground swelling, which can be detected by the global positioning system (GPS) and space satellites; radon gas found in groundwater; underground rock movements detected by lasers; and unusual animal behavior.

On the basis of several of these precursors, Chinese geologists warned of a probable earthquake in northeast China in 1975, which precipitated an evacuation of several million people. Nine and a half hours later, a great earthquake struck the area and tens of thousands of lives were saved. Unfortunately, a year later, another large earthquake struck with no precursory signs. Modern seismic monitoring networks have made earthquake forecasts and warnings more accurate, but earthquake prediction is still far from perfect.

For this reason, disaster prevention efforts focus heavily on earthquake preparedness in addition to prediction. In the United States, the earthquake-prone state of California has taken the lead in requiring earthquake-resistant buildings and bridges and crafting emergency-response procedures. Earthquakes are inevitable, but through science, technology, and responsible leadership, the death and destruction associated with earthquakes can be reduced.

The Science and Study of Earthquakes

What Causes Earthquakes

By Stephen L. Harris

In this selection, geology enthusiast Stephen L. Harris discusses how an earthquake works, with specific reference to the San Andreas Fault in California where two huge tectonic plates, the Pacific plate and the North American plate, are rubbing against each other. Harris describes the different types of earthquake waves and the damage they can cause. He further explains the process of liquefaction, where soil and sand are changed into a liquid state, causing some of the worst earthquake damage as saturated sands shake violently and heavy buildings tilt or sink into the liquid ground. In addition to violent shaking, earthquakes cause damage through landslides and seismic sea waves.

Stephen L. Harris is a professor of humanities at California State University in Sacramento. His avocation is geology and he has published three books and a series of articles on the subject.

An earthquake is the sudden trembling or shaking of the ground caused by the abrupt movement or displacement of rock masses within the earth's crust. Most earthquakes originate within the upper ten to twenty miles of the lithosphere, the earth's rigid outer shell. Powerful forces in the lithosphere exert stress on the rock, pushing or pulling it. Rock is elastic enough to accumulate strain, bending or changing shape and volume. When stress exceeds the strength of the rock, the rock breaks along a preexisting or new fracture plane called a fault. The fracture rapidly extends outward from its place of origin, the focus. As the rock breaks, waves of energy—seismic waves—radiate through the earth, causing the vibrating and shaking of an earthquake.

Seismic waves, generated by friction and crushing as masses of rock slide past one another, travel outward from the earthquake

Stephen L. Harris, *Agents of Chaos: Earthquakes, Volcanoes, and Other Natural Disasters.* Missoula, MT: Mountain Press Publishing Company, 1990. Copyright © 1990 by Stephen L. Harris. Reproduced by permission of the publisher.

focus like ripples on a pond. The fastest are the primary, or P, waves, which compress the rock in front of them and elongate it behind as they rush through the planet at about three to four miles per second. Next come the S, or secondary, waves, which undulate, causing an up-and-down and side-to-side motion as they roll through at about two miles per second.

Most damaging to man-made structures are the surface waves, called Rayleigh and Love waves, which shake the ground both vertically and horizontally. They can create a visible roiling and billowing of the surface as well as a jerky zig-zag motion that is particularly destructive to high-rise buildings. The slowest moving of seismic waves, surface waves cause the worst devastation because they accelerate ground motion and take longer to travel through a given area.

Damage from Quakes

The degree of damage to buildings and landscapes depends largely on their proximity to the epicenter and the nature of the underlying soil. In general, damage is most severe within a few tens of miles of the earthquake source and diminishes with increasing distance. Certain kinds of soil, however, increase the intensity of destructive shaking, even many tens of miles from the

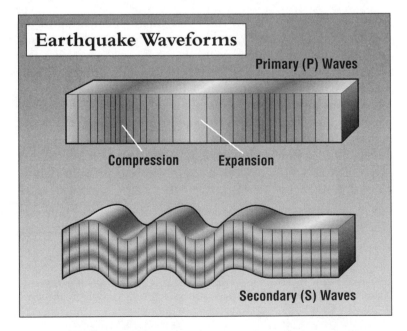

Earthquake Waveforms

Primary (P) Waves

Compression   Expansion

Secondary (S) Waves

epicenter. In 1989 San Francisco's Marina District, although nearly sixty miles from the point at which the San Andreas fault ruptured, suffered extensive damage. The Marina soils, sandy landfill covering an old lagoon, underwent partial liquefaction. A secondary effect of earthquakes, liquefaction occurs when water-saturated sands or silts shake violently; water pressure forces sand grains apart, causing the subsurface materials to act as a liquid. Transformed into mush, liquified sediment can flow laterally underground, cracking the ground surface and allowing heavy buildings to tilt or sink into the liquified muck. The syrupy soil may break in waves, as it did in the Marina, or erupt in geysers of liquid sand. The liquefaction of unconsolidated soils with a high water table is responsible for some of the worst damage in many earthquakes, including those that hit Alaska in 1964, Mexico City in 1985, and the Bay Area in 1989.

The San Andreas Fault

America's most celebrated source of large earthquakes, the San Andreas fault slices through California's coastal mountains from the Mexican border to Cape Mendocino, where it passes into the ocean floor. The San Andreas marks the boundary between two immense slabs of crustal rock that are slowly grinding past each other. The slab or tectonic plate holding the Pacific Ocean basin creeps northwestward at the rate of an inch or two per year, rubbing against the western edge of the North American plate. When the two crustal plates become temporarily locked together, strain accumulates until the plates break apart, releasing seismic energy that sets the Earth vibrating. Large-scale movement along two different sections of the 700-mile-long San Andreas fault has triggered two of the greatest earthquakes in United States history, those of 1857 and 1906. The east and west sides of the San Andreas changed their relative positions only a few feet during the 1989 quake, but in 1906 they shifted as much as twenty-two feet.

The underground rupture initiating an earthquake will continue until it reaches areas in which the rock is not sufficiently strained to permit it to extend farther. The 1906 earthquake ruptured the northern San Andreas for a distance of 280 miles, most of the length marked by visible surface fracturing. Despite considerable secondary cracking of the ground surface in the Santa Cruz Mountains and elsewhere, the 1989 quake did not produce any detectable primary surface rupture.

Besides causing damage through fault movement, violent shaking, soil liquefaction, ground failure, and landslides, earthquakes may also trigger tsunamis, seismic sea waves that can travel thousands of miles across oceans at speeds of 300 to 400 miles an hour. As the tsunamis approach shore, they begin to drag on the seafloor, slow down, and rise to heights of fifty feet or more. Such earthquake-produced waves have repeatedly slammed into low-lying coastal areas in Hawaii, Alaska, and the Pacific Northwest.

The numerous earthquakes and dozens of volcanic eruptions that occur annually around the Pacific rim are not unusual. These bursts of violence that terrorize people in Tokyo, Tacoma, San Francisco, or Mexico City are merely demonstrations that earth's processes operate chaotically, pulsating with sudden change and movement.

Charles Richter
and the
Earthquake Scale

By Caroline Green

Today the science of earthquakes has newer and better ways to mea-sure earthquakes, but Charles Richter laid the groundwork in the 1930s with a good basic system for earthquake measurement. In this article, freelance science journalist Caroline Green relates a brief life history of Richter, a brilliant scientist born in 1900. Green explains that Richter was idiosyncratic as well as obsessed with the science of earthquakes to the point of keeping a seismograph in his living room with printed read-outs continuously draped over the furniture. Richter and fellow scientist Beno Gutenberg devised the now famous Richter scale that measures earthquake magnitude, indicating the energy released in an earthquake. Richter was also concerned with improving structural standards in build-ings to minimize earthquake damage. He continued his deep immersion in the field of seismology until his death in 1985.

Horrific pictures were beamed around the world in Jan-uary 1995 as rescue workers in Kobe, Japan, frantically searched through wreckage for the survivors of a mas-sive earthquake that had ripped the city apart. It was the fourth earthquake to hit Japan in less than a month, but it was the most destructive and deadly that the country had witnessed for many years. News reports differed around the world, but each one in-cluded the same statistic: that the earthquake had measured 7.2 on the Richter Scale.

When seismologist Charles Richter, who died in 1985, found a way to measure the magnitude of earthquakes in the 1930s, he

Caroline Green, "The Man Who Loved Earthquakes," *Geographical Magazine*, vol. 67, December 1995, p. 34. Copyright © 1995 by *Geographical Magazine*. Reproduced by permission.

became a household name. The science of seismology may be a mystery to the vast majority of people, but most know that the Richter Scale relates to earthquakes, and that the higher the rating, the bigger the quake.

There are a variety of ways to measure earthquakes these days. But, much to the irritation of the media, they usually all get lumped by the media under the banner of the Richter Scale. "The Richter Scale is still used, but most seismologists have newer and better ways of measuring earthquakes now," says Dr Wayne Richardson, a senior seismologist at the International Seismological Centre based at That cham, near Newbury, Berkshire. "However, Richter produced a nice, simple method that formed the basis of them all."

Early attempts to quantify earthquakes used a concept called 'intensity'. This is an assessment of the strength of shaking caused by the earthquake based on its effects—for example the amount of damage caused to buildings, and public distress—and is therefore partly subjective. Various scales were proposed by different scientists; the first to gain widespread international acceptance was that produced by Michele de Rossi and Francoise-Alphonse Forel in the 1880s, known as the Rossi-Forel scale. This was subjected to various modifications, notably by Guiseppe Mercalli, whose name has been attached to a version of the scale.

Intensity is a useful concept, still much studied today, but assessing the size of an earthquake on its own is not enough, partly because it varies from place to place, being greatest near the epicentre and much less further away. If the epicentre happens to be in the sea, then intensity becomes hard to use.

A Good Basic Scale

During the 1930s, Charles Richter and Beno Gutenberg began to look at ways to classify earthquakes at the California Institute of Technology (CIT) into small, medium and large, using the instrumental records from seismographs. Richter and Gutenberg reasoned that if two earthquakes occur at the same distance from an observer, the one that produces a larger amplitude of ground motion can be considered to be the larger earthquake. But since the amplitude of ground motion decreases the further away the observer is from the epicentre, distance must also be taken into account. Working with data recorded in southern California at a number of seismograph stations—all of which used the Wood-

Anderson seismometer—Richter plotted graphs showing the decay of ground motion amplitude with distance. By taking the response of the seismometer to one arbitrarily chosen small earthquake at a certain distance, and giving it the value zero, he could then assign values to other earthquakes according to the ground amplitudes they produced, when corrected for distance. Each step up in the scale represents a tenfold increase in the amplitude of ground motion. In energy terms, each step in the scale represents an increase of roughly 30.

It was a good basic system, which made it simple to calculate the magnitude of any earthquake measured on a seismograph. The problem was that it was really only applicable in southern California, where earthquakes tend to be quite shallow. In Japan, where seismic activity tends to be much greater and earthquakes are generally much deeper, anomalies in the calculations arose. Neither was it very helpful for measuring long-term seismic activity. "The scale was such a success that Richter almost regretted it," says Wayne Richardson. "It made the science look too simple." Nonetheless, important groundwork was laid for the accurate measurement of earthquakes, securing Richter a rightful place in history.

Beno Gutenberg's role was played up in later years by Richter. "The usual designation of the magnitude scale in my name," he said in 1977, "though perhaps convenient, does less than justice to the great part played by Dr Gutenberg." But according to a colleague quoted in Richter's obituary in the *Los Angeles Times*: "For many, many years, Charlie did very little to emphasise Beno's role. If you wanted to think it had all been Richter's doing, then that was all right with Charlie."

Critics said Richter was fascinated with earthquakes to the point of obsession. He kept a seismograph in his living room and visitors would find printed readouts permanently draped across the furniture. Richter used to say his wife liked them there as a conversation piece. He adored talking to the press about the latest earthquake and, when one occurred, he used to clutch the seismology lab telephone in his lap so that he could be the first to answer the calls when they came in.

Richter's Life

Charles Francis Richter was born in April 1900 on a farm near Cincinnati, Ohio. He was raised by his maternal grandmother and, when he was nine, the family moved to Los Angeles. A year later, Richter experienced his first earthquake, which he said "surprised [him] no end". He obtained a degree in physics from Stanford University before enrolling at the CIT for graduate study in 1920. In what was only supposed to be a temporary assignment, Richter began work in the fledgling seismology laboratory, where he stayed for virtually his entire career.

Over the years, he developed a reputation for being a deeply gifted but idiosyncratic man. "He had a brilliant mind, but it bordered on instability," said one colleague after Richter died. He married his wife Lillian in 1928 and the couple never had any children. They were thought of as an unusual, if contented, pair. To begin with, they were nudists—which was pretty shocking stuff in those days. But despite this dating behaviour, Richter hated socialising and had little time for the departmental get-togethers that are such a big part of academic life.

He remained totally immersed in his subject right up until his death, attending lectures into his 80s. Like many seismologists, Richter did not spend much time trying to predict earthquakes. As Wayne Richardson says: "The only thing you can really predict is that they are going to keep happening, but that's about it. Some say that accurate prediction is impossible."

One of Richter's chief concerns was to improve structural standards in buildings in southern California to minimise the effects of earthquakes. The Maharashtra earthquake in India in 1993 killed 10,000 people, while the Californian earthquake of January 1994, which was nearly twice as powerful, killed just 40. The difference in fatalities was mainly due to the stability of the buildings affected by the earthquake.

Every time a major earthquake strikes, a little more is learned about how to survive the next one. But one thing is certain: "We are becoming more at risk," says Wayne Richardson. His team records 5,000 earthquakes a month worldwide, including the smallest tremors. Their frequency is not rising but as the world's population grows by 93 million each year, and cities become more crowded—45 per cent of the world's population now live in towns and cities—so the risks increase. The science that so gripped the strange, brilliant mind of Charles Richter still has a long way to go.

Large Earthquakes Trigger Distant Quakes

By Richard A. Kerr

Seismologists have been debating a relatively new stress-wave theory of earthquake causation. Richard A. Kerr, a staff writer for Science *magazine, explains that seismologists have long known that earthquakes create seismic waves that affect neighboring faults, generating further quake activity. The new earthquake stress-wave theory suggests that seismic waves travel much slower and extend farther than earlier estimates. Using computer simulations, a group of geophysicists demonstrated that several large earthquakes in the 1950s and '60s may have caused some of the seismic activity of the '70s and '80s. Other scientists, unconvinced of the theory's validity, believe that the correlations may only be coincidences.*

E arthquakes were once thought to keep to themselves, striking on a schedule determined only by the history of each particular fault. Then seismologists began to realize that every rupturing fault communicates with neighboring faults, instantly reaching out tens or hundreds of kilometers to hasten or delay distant earthquakes. Now a group of geophysicists suggests that these lines of communication extend even farther— and carry much, much slower messages.

Big quakes can trigger other quakes thousands of kilometers away and decades later, according to calculations presented in *Science*. Geophysicists Fred F. Pollitz and Roland Burgmann of the University of California, Davis, and seismologist Barbara Romanowicz of UC Berkeley simulated how stress travels through deep, viscous rock. They found that the great earthquakes that struck the far North Pacific in the 1950s and '60s could have set

Richard A. Kerr, "Can Great Quakes Extend Their Reach?" *Science*, vol. 280, May 1998, p. 1,194. Copyright © 1998 by American Association for the Advancement of Science. Reproduced by permission.

off waves that triggered a pulse of seismic activity in California in the 1980s.

"It's an exciting possibility," says seismologist Thomas Hanks of the U.S. Geological Survey (USGS) in Menlo Park, California. If the reach of big quakes extends that far, seismologists may be able to make more sense of the comings and goings of earthquakes worldwide, he adds. Researchers are intrigued, though not yet completely convinced. "It could be right," says tectonophysicist Wayne Thatcher of the USGS, "but I think it has a ways to go before being a persuasive argument."

Researchers have long recognized a potential transmission route for long-distance messages among faults: the thin layer of soft rock at depths of 80 kilometers or more called the asthenosphere. There, temperature and pressure combine to soften rock so that, although still solid, it slowly flows over decades. The more rigid tectonic plates that make up Earth's surface, such as the great Pacific Plate, glide along on this softer layer.

But plates don't slide smoothly at their edges. They stick to each other, build up stress, and then jerk forward in earthquakes. The quake redistributes stress nearby, adding stress in some places and relieving it in others. For example, between 1952 and 1965, four great quakes struck along the Aleutians and the Kamchatka Peninsula, where the Pacific Plate is diving beneath the North American Plate. After each quake, the Pacific Plate adjusted to the new plate positions immediately, stretching like a sheet of rubber and triggering flow in the asthenosphere below. Spreading outward through the asthenosphere like the ripple of a pebble dropped in a pond, the wave created by this flow could transmit the stress induced by the quakes.

"That [stress] wave has to exist," says Burgmann. "The only question is how strong is it?" To find out, the group created a computer simulation of elastic plates, ductile asthenosphere, and large earthquakes in the northern North Pacific. In the model, the stress wave generated by the quakes moved southward across the Pacific and northward under the Arctic Ocean at a rate that depended on the viscosity of the asthenosphere. When researchers plugged in a viscosity that Romanowicz calls "reasonable but a bit on the low side" of current estimates, the crest of the stress wave entered the eastern Arctic Ocean in the 1970s; it passed off British Columbia around 1975, and California around 1985. Wherever the wave passed, it briefly accelerated plate motions, which could have spurred earthquake activity.

The timing is a good fit to surges of seismic activity, say Pollitz and his colleagues. According to the model, the wave may have triggered the surge of magnitude 5 and greater quakes observed in the eastern Arctic Basin in the 1980s. To the south, the wave's progress—marked by accelerations of only a couple of millimeters per year—could be seen in pulses of increased seismicity in Northern California in the 1970s and Southern California in the 1980s.

Even the types of earthquakes seemed to fit stress-wave triggering, says the group. The Southern California seismicity mostly took the form of quakes on faults other than the San Andreas. The sides of these faults move chiefly up and down rather than sideways, as the San Andreas does. That feature of the seismicity was noted in 1995 by seismologists Frank Press of the Washington Advisory Group in Washington, D.C., and Clarence Allen of the California Institute of Technology in Pasadena, who speculated that a stress wave oriented to favor vertical fault motions might be responsible. The wave set off by the great Alaskan quakes fits the bill, Pollitz's team says. "The whole thing seems to hang together," says Press.

But others point out that the correlation of the passing wave with a flurry of seismicity could be chance. "There have been many interesting patterns in seismology that have turned out to be wonderful coincidences," says seismologist Lucile Jones of the USGS in Pasadena. And the stress wave, dampened by distance, seems too weak to trigger quakes, other researchers say. The extra strain added by the wave would be "really small," says Thatcher, perhaps a factor of 10 smaller than what's assumed to trigger quakes at short range.

"Yes, the strain changes are small," concedes Romanowicz, "but I don't think anyone has a definite idea of how much strain you need to trigger an earthquake." Burgmann adds that a coincidence is unlikely, because seismicity along the entire west coast fits the stress-wave theory.

Further tests of this bold idea are in the works. Although existing geodetic networks weren't sensitive enough to catch the subtle stress changes signaling the arrival of the wave, says Burgmann, current systems should record its departure in the coming decade. Then it should become clear whether distant earthquakes are the interfering busybodies he and his colleagues suspect they are.

Tracking Earthquakes in Africa

By Christopher Scholz

Geologist Christopher Scholz, hired to act as an earthquake consultant in the south African country of Botswana, describes his "safari," not to hunt animals, but in search of knowledge about earthquakes. In addition to recording earthquakes for future developmental purposes in the area, Scholz, a technician, and eleven crew members were researching how continental crusts rift, or stretch, to form faults. Scholz installed a network of seismic stations in what he calls a microearthquake survey. This survey studied many small, frequent earthquakes and gleaned as much data from them as from the much less frequent larger quakes. As Scholz describes, the trip had its adventures, including tense encounters with elephants and local Bushmen.

Teddy and I were sitting about twenty yards apart. We had been like that for more than an hour, hunched up against the trunks of a couple of mopani trees as we waited for the herd of elephants to leave the grove we were in. They had moved into the grove from several directions, and by the time we had noticed them, we had lost any chance of retreating to the Land Rover. There was nothing for us to do in the meantime but wait them out and to try to be as inconspicuous as possible.

Our big fear was that one of the small grazing groups would select one of our trees and be startled by our presence. There wasn't much we could do in that case. You can't outrun a charging elephant who might want to trample you. One time, a mother elephant came close enough to catch our scent in the still

Christopher Scholz, *Fieldwork: A Geologist's Memoir of the Kalahari*. Princeton, NJ: Princeton University Press, 1997. Copyright © 1997 by Christopher Scholz. Reproduced by permission.

air, and snorting, she wheeled away, whisking her calf in front of her with her long trunk. It was a close call.

Climbing a tree would have been no refuge in this situation. That offers protection from Cape buffalo, but not from elephants, which can reach the upper branches of trees with their trunks. And the elephants were not showing any indication of leaving any time soon. The trees were full of succulent fruit, and the elephants were munching away contentedly, pulling whole branches down for their young.

We didn't have access to water, and thirst was starting to claw at my throat. With dry lips I gazed desirously at some half-rotten pieces of fruit lying on the ground ten feet away, but I didn't dare try to go get them.

After about two hours, trumpeting rang out from the south. The elephants near us perked up their ears and listened a moment, then headed off in a brisk trot in the direction of the trumpeting. They had been summoned for their afternoon trek to water. "Lunch break's over. Everybody back on the bus."

Finding ourselves alone at last in the grove, Teddy and I stood up and looked around warily. When we had convinced ourselves that there were no stragglers left, we started hiking back to where we had left the Land Rover. "One thing I can say about you, Scholz," said Teddy, "you sure can pick the places to go to study earthquakes."

Studying Earthquakes in Botswana

It was the spring of 1974 in southern Africa, and I had my reasons for being out there trying to record earthquakes in the hot sun. The United Nations Food and Agricultural Organization had asked me to serve as an earthquake consultant in Botswana, a country about the size of France located just north of South Africa. For future development purposes, the organization was making a broad ecological study of the Great Okavango Swamps, the only source of fresh surface water in a country consisting mostly of the Kalahari Desert.

At the time, I had no idea that Botswana was seismically active. To my knowledge, the only active region in Africa was the East African Rift System, a zone where Earth's crust is pulling apart. I thought this system was limited to a region far to the north. But after searching through seismic records and earthquake catalogs of southern Africa, I found that two magnitude 6.5

quakes had occurred in the Okavango in the early 1950s.

More evidence came from satellite images that showed linear features in the land running southwest across Botswana. There are very few features in nature that are really straight, especially at this scale of tens to hundreds of miles. They could have been linear sand dunes, but they were running parallel to a pattern of seismic activity plotted on other maps of the area. That made me think they were tectonic faults. If they were, then the most likely explanation was that I was looking at the tip of a previously unrecognized branch of the African rift system.

To prove the existence of a previously unknown branch of the African rifts was a nice little scientific problem. But a grander problem loomed. The theory of plate tectonics has not yet succeeded in solving the root problem of the cause and mechanism of rifting, which often takes place within a single plate. In one class of models, rifting is initiated by hot spots, where plumes of magma rising from the deep mantle push up and stretch the crust. But Botswana was too far south of the hot spot under the East African Rift System to have been influenced by the forces associated with it. There was a chance, a slight chance, that this little region of northern Botswana could help answer the question of how continents rift.

Microearthquake Survey

What I had in mind was something called a microearthquake survey. This technique takes advantage of one of the few empirical laws of seismology, called the Gutenberg-Richter relation. According to this law, in any seismically active region there are always many more small earthquakes than large ones, following a very regular pattern. The frequent but small earthquakes carry the same information as the large but infrequent ones. So the idea of a microearthquake survey is to install a local network of very sensitive seismic stations capable of recording magnitude 1 and 2 earthquakes. This way you can collect data in a few months that might normally take you forty or fifty years, were you to survey only larger earthquakes.

I wasn't in the field alone. I had my technician Teddy Koczynski and eleven crew members. We were going to run our own safari, but we were after knowledge instead of animals. In a place like this, that would be a venture combining the most exalted and the most basic of human challenges. We had to solve a scientific problem, and we also had to survive.

We met Teddy and his crew lounging under a big tree on the road back to Toteng, which is just south of the Okavango Delta. As we pulled up, Teddy announced, with a big grin, "Hey, we found Toteng Star."

Toteng Star

Toteng Star was a nearby rock outcrop that was indicated with a star on the geologic map of Botswana. We had been looking for the outcrop for a couple of days because it was a possible bedrock site for our seismic instruments. Seismic waves travel more effectively through dense bedrock than through the loose Kalihari sand, where the small signals we were looking for might be lost.

"Phew, what a relief. I was beginning to wonder if it really existed."

"It isn't all that big. You won't believe it. You could have driven right over it in the dark. We wouldn't have found it except for the cows."

"The cows?"

"Yeah, there were cows standing all around it. It must be salty, or something. Come on, check it out."

We followed Teddy's Land Rover along a track that threaded its way through a maze of scrub, finally coming out upon a barren flat occupied by a desultory-looking herd of cattle. Beyond them was that rarest of sights in this desert: a low outcrop of rock. We got out and walked around it. I hit it with my hammer and it rang like a bell. This was no loose boulder. No more than two feet high and ten feet across, it was a solid outcrop of peridotite, a chunk of two-billion-year-old mantle that was directly connected to the seismic activity deep below the surface. It was a piece of honest-to-goodness bedrock poking its lovely nose out of the desert floor.

Teddy had set up the instrument on the far side of the outcrop. I thumped lightly on the rock with my foot, and sharp signals appeared on the recording drum. This site was about six hundred times more sensitive than any instrument we could run in the sand. If there was any microearthquake activity at all in this vicinity, we would soon know about it. We finally had a big ear to the ground.

These suspicions were confirmed when we checked the Toteng Star instrument the next day. It had recorded three microearthquakes overnight. By contrast, the instruments sited in sand had recorded only one such event in the previous five days. We knew that in order to locate more of these tiny earthquakes, we had to set up instruments at the other rock sites, and as soon as possible. Every day they weren't operating meant another day of lost data.

Bushman Legends

That evening Teddy and I went over our options. Two other sites looked like good possibilities. One was an outcrop east of Lake Ngami near a U.S. Steel camp. The third was the Kgwebe Hills, rocky hills called "kopjes" that poke out of the Kalahari at irregular intervals. We knew full well where the Kgwebe Hills were: fewer than fifty miles away, as the crow flies. The problem was that we didn't know how to get there, and our guide, Thomas, was not very forthcoming in that regard.

"Ted, we've got a serious problem. We have to get to the Kgwebe Hills, but something's fishy about that whole subject. Ever since we moved down here to Toteng, I've been bugging Thomas about going there, but all he ever comes up with is something like 'you can't get there from here.'"

"Yeah, I know . . . I get the feeling that the crew doesn't want to go there. It seems like they're scared to death of going near the kopjes because they are supposed to be Bushman holy places, or something."

"Ah, so that's it. Stories about the Bushmen and the kopjes are famous legends in this part of the world. Laurens van der Post wrote about it in *Lost World of the Kalahari*. According to him, the Bushmen believe their gods live in the kopjes, and the Tswana, who are scared of the Bushmen anyway, think the Bushman spirits haunt them."

"What are these Bushman kopje stories? Just curious."

"Oh, come on, Teddy. It's just a lot of hokum. Van der Post told this story that they couldn't film the old Bushman petroglyphs in the Tsodilo Hills, supposedly because they were protected by a nest of cobras. Then they got chased out of the kopjes by a swarm of bees. He tried to make out that these were some kind of Bushman spirits or something. Hey, it added some zip to his book. You've gotta sell copies somehow."

We headed out of camp the next day in a caravan of both Land Rovers and crew. The plan was to follow the main road south all the way to Ghanzi looking for any well-trafficked route to the east. Eventually we found our second site: trenches dug by the U.S. Steel camp that reached down into rock. Unfortunately, the rock in the trenches turned out to be pretty broken up and crumbly, nothing compared to the rock quality at Toteng Star. But it was still a far better site than any on sand. From there we could just see the Kgwebe Hills looming on the horizon to the northeast, but since it was late in the day, I decided to postpone that visit until the morning.

After dinner, Thomas came into our camp. This was something he had never done before; it was a breach of camp etiquette. Normally the crew stayed in their own camp during off hours.

He stood in the shadows a few feet from our mess table. "We can't go to Kgwebe Hills, boss. It's too dangerous."

"What do you mean 'dangerous,' Thomas?"

"Kgwebe Hills is Bushman place, boss. Too dangerous for us to visit. Da crew, dey very scared of dat place."

"Well, maybe they are, Thomas. Or maybe they've been listening to too many stories. But we are going to the Kgwebe Hills tomorrow. If the crew is afraid to come, fine; they can stay in camp. Teddy and I will go by ourselves. We know the way."

I can't say that I was completely without qualms at that point. I was not concerned with the superstitions of the crew, but I had grown accustomed to following Thomas's advice regarding bush lore, and I didn't feel too comfortable in disregarding it this time, no matter how ridiculous it seemed to be.

Visiting the Kgwebe Hills

By eight the next morning our skeptical crew had decided to come along, and we headed out for the Kgwebe Hills. The road had been graded at one time and was still in pretty good shape. The hills themselves appeared as gray lumps on the horizon. As we approached them, Thomas and the crew, who normally chattered throughout the day, fell silent. We continued to ascend the gentle slope to the base of the kopjes, which seemed to rise out of the desert floor like immense, shiny gray dumplings. Baobab trees and leafier, more succulent desert shrubs began to appear, a sure sign of rock and a water table at shallow depth.

At the top of the incline, the road suddenly spilled out into a valley where we came to a halt. It was like a Shangri-la: a circular, flat-bottomed valley maybe a mile across, surrounded by the kopjes on three sides and green with vegetation. On the far side of the valley there were leafy green trees, like banana trees. We could also just make out what might be the brown thatch of hut roofs here and there among the trees. There were no other signs of habitation. Teddy, Thomas, and I descended from the Land Rovers and surveyed this peaceful scene. The other driver and the crew, frightened beyond belief, stayed inside the vehicles. In front of us were two shallow circular pits that looked as if they might have once been deeper but had slumped in. They certainly didn't look natural. We walked over to examine them. Thomas, following, said, "Those are da pits to catch da animals. Da Old Ones make dem, long time ago. Den dey herd da animals into da valley and push 'em dis way. Da animals, dey try to get away, dis way!" he said, gesturing to the road we had come in on, which followed a narrow break in the rocky cliffs. "Den de animals fall in de pits."

"Oh. So these were animal traps made by Bushmen?"

"No, not by Bushmen. By da Old Ones. Before da Bushmen come."

The crew were too spooked to work, so Teddy and I unloaded the gear ourselves, and hauled it a hundred yards or so around the perimeter of the big kopje to the left of the road. The rock rose straight from the sand at a steep angle, with just a few rounded boulders forming a rubble pile at its base. We tested the rock here and there with our hammers. It also was of that pure, bell-like quality that we had found at Toteng Star. This was going to be a site of equally superb seismic sensitivity.

We installed the geophone, a sensor that picks up deep-Earth rumblings. I fully battened everything down, burying the cables with large rocks and building strong shelters for the geophones, batteries, and recorder as extra precautions against baboons that might live among the rocks.

Bushmen Encounter

When we were very satisfied with our installation, we turned and walked back to the Land Rovers. Then . . . "Oh, my God!" I received a shock like a chemical rush. It was actually accompanied by the sensation of the hair standing up on the nape of my neck. Teddy and I stood stock still, staring at the sight before us.

Thomas was standing in the clearing in front of the Land

Rovers, talking animatedly, in the clicking tongue of the Bushmen, to what looked like a Bushman chief, behind whom, on a low rise, were squatting some forty Bushmen in two rows, clad only in loincloths and holding spears upright, butt-ends on the ground. They were staring straight ahead at Thomas and the Land Rovers with fixed and unblinking expressions, muscles bunched and ready for action.

I walked up to Thomas, but before I could utter a word, he said, out of the corner of his mouth, "Get in the Land Rover." I did what he said. Teddy and I sat in our Land Rovers, watching Thomas continuing the negotiations, not understanding a word of what was being said. In back, the crew were lying with their arms covering their heads, whimpering. The Bushmen, in their ranks, didn't move a muscle. Finally Thomas returned, climbed in, and we started up. We turned around and drove back down the road, and a shroud of anxiety gave way to a sense of relief. Not a word was spoken all the way back to camp, each of us being left to our private thoughts.

Analyzing the Data

After a month or so of rest and recuperation at home, I sat down to analyze the data. Our instrument recordings showed a band of seismic activity running northeasterly from just south of Lake Ngami all the way to the Zambezi River, following activity over to the east and joining up with activity in the Kariba Gorge. Combining these results with data obtained by the Rhodesian Seismic Network, I was able to show this was the terminus of an arm of the African rift system extending from the Lwangwa Valley down to the Kalahari. And because there were no volcanoes in this part of the country, or other deformities that often occur in the land before it rifts, I was also able to show that a rift could extend across relatively flat land.

In the ensuing twenty years, there has been little or no progress on the main problem: the cause and mechanism of continental rifting. Although this is still a big puzzle in geology, it is a stagnant one. In the absence of a breakthrough, it is a problem that has been tabled, with people working just on the fringes. If and when new progress is made on continental rifting, our published work may enjoy a brief day in the sun. But the Botswana fieldwork was about much more than science or career. Some things in life are worth doing solely for the experience.

Monitoring Quakes from Space: The Global Positioning System

By Dan Hogan

The global positioning system (GPS), which uses radio signals trans-mitted by twenty-four satellites orbiting the earth, is being used by scien-tists to measure movements of the earth's crust that even highly sensitive seismographs can miss. In the following selection, Dan Hogan explains that the GPS network is capable of measuring movements of the earth's crust as small as one millimeter per year. By monitoring where tectonic strain is increasing, scientists hope to improve earthquake predictions. The GPS system can also detect underground or "stealth quakes," which never break the earth's surface but can be as powerful as the ones that do. Dan Hogan is a science writer and former managing editor for Current Science.

Residents of Northridge, Calif., may not know it, but the earthquake that devastated their community and parts of nearby Los Angeles in 1994 is still rumbling under their feet today as a silent, hidden "stealth quake."

Scientists made the surprising discovery recently after they analyzed data from new earthquake-monitoring devices set up throughout southern California. Instead of measuring shaking in the ground, as traditional earthquake instruments called seis-mographs do, the new devices measure tiny movements in the ground using radio signals transmitted by 24 satellites orbiting Earth.

Those satellites make up what is called the Global Positioning System, or GPS for short. GPS was originally designed by the United States military to locate tanks, planes, and ships. The system now has many other uses, such as helping boaters navigate unknown waters and guiding lost hikers back to their camps. Using a hand-held device called a GPS receiver, a person can find his or her exact latitude, longitude, and elevation at any time and any place on Earth.

Earthquake scientists, called seismologists, are also using GPS receivers now. The devices help them measure movements in the ground that are usually too slow and weak to detect. Even highly sensitive seismographs can miss those tiny movements. For example, GPS receivers have detected a steady upward movement of the ground in Northridge. That movement has caused the nearby Granada Hills to rise 15 centimeters (6 inches) in the past three years.

What Causes Earthquakes

Earthquakes occur when plates—huge sections of Earth's crust that float on the planet's hot, molten interior—slide past one another, collide head-on, or separate from one another. Over time, the movement of Earth's plates can cause strain to build up along plate boundaries called faults.

When that strain becomes too great, the plates move suddenly, releasing the energy stored in the strain. That released energy often sets off violent shaking that begins at points along faults called epicenters. The vibrations from the shaking spread out from the epicenters in the form of tremors called seismic waves that travel through the ground.

Sometimes that sudden movement of Earth's plates is so great that faults break through the surface, forming visible fractures in the ground. Those fractures range in size from small cracks in roads and sidewalks to enormous gashes, such as the San Andreas fault, which runs two-thirds of the length of California.

"Stealth Quakes"

However, many faults lie buried in the ground and cannot be seen from above. "The Northridge quake occurred on a fault that did not break all the way to the surface," explains Gregory Lyzenga, a geophysicist at Harvey Mudd College in Claremont, Calif.

"However, the layers of rock just below the surface near the epicenter have continued to move in a fluidlike manner—sort of like honey flowing off a spoon—since the earthquake."

The gradual movement from that "stealth quake," as Lyzenga calls it, is about the same amount that would occur in an earthquake with a magnitude of 6.0. Scientists determine the magnitude, or size, of an earthquake according to the amount of ground movement, the speed of the seismic waves, and other measurements. The 1994 Northridge earthquake, which was considered strong, had a magnitude of 6.7.

Using the new earthquake-monitoring GPS receivers, scientists plan to gather data in California and elsewhere in the United States from other stealth quakes that have gone undetected until now.

"The GPS network will continuously measure movements of the Earth's crust with a precision of 1 millimeter per year, which will show us where strain is building up," says Andrea Donnellan, a scientist at the National Aeronautics and Space Administration's Jet Propulsion Laboratory in Pasadena, Calif.

By understanding where that strain is building up, scientists hope they'll have a better chance of predicting earthquakes and saving lives and property in the future.

How the GPS Works

How does the Global Positioning System work? GPS satellites continuously broadcast radio messages. Each message contains a very accurate time signal, a rough estimate of the satellite's position in space, and a set of coded information that a GPS receiver on the ground can decipher.

A receiver uses its internal clock and the coded information from GPS satellites to tell the time it takes for the GPS signals to reach the receiver. Using that information, the receiver can calculate the distance to each satellite. Once the receiver knows the distances to at least four satellites and their positions, the receiver can determine a person's position on Earth.

By comparing GPS measurements taken at different times from receivers set up at different locations, earthquake scientists can see the amount of shift in Earth's crust between those locations over time.

Earthquake Disasters

The Great San Francisco Earthquake

By Kerry Sieh and Simon LeVay

The Great San Francisco Earthquake, which struck in the early morning on April 18, 1906, was one of the largest earthquakes ever to hit California. The property damage was estimated to be about $7 billion in current dollars. Although the earthquake itself lasted only about a minute, the fires that broke out in the quake's aftershocks raged for three days, destroying much of the city. Although close to five hundred people were found dead, four times that many were unaccounted for and probably perished in the quake and ensuing fires. Statewide, San Francisco suffered most of the destruction, but even more intense shaking was felt in Santa Rosa, Stanford University, and San Jose. People felt the earthquake over an area of one hundred seventy-five thousand square miles.

The quake was also significant for scientific reasons. It gave clues to the nature of faulting and earthquakes in general. The exhaustive studies done following the quake provided needed research data, a notable contribution to modern seismology.

Kerry Sieh is a professor of geology in the Seismology Laboratory at the California Institute of Technology and author of many scientific articles on earthquake geology. Simon LeVay is a well-known science writer and neuroscientist who has taught at Harvard Medical School and the Salk Institute for Biological Studies.

The infamous earthquake struck at 12 minutes after five in the morning. Many agricultural workers were already in the fields; some of them heard a rumbling sound, and then, by the gray predawn light, they saw waves riding along the ground as if it had suddenly been transformed into a windswept sea. Some trees were shaken so severely that their crowns touched the ground, and the trunks of others snapped off entirely. Animals panicked; riders were thrown from their horses and cattle fell to the ground. Prodigious quantities of milk were spilled from churns. Water tanks overturned, and the ground itself failed in many places, torn by landslides, fissures, and sand-blows. Railroad lines buckled and water pipes broke, sometimes in hundreds of places. The shaking lasted about a minute.

The earthquake was dramatic in the countryside, but in the cities it was catastrophic. Many of the inhabitants of San Francisco, especially the well-to-do, were still asleep at the time, but they were rudely awakened by the first shock. The great Italian tenor, Enrico Caruso, who had had a triumphant success in *Carmen* the previous evening, rushed down to the lobby of the Palace Hotel, where he threw a tantrum worthy of the star he was, weeping floods of tears and swearing never to revisit a city that permitted such disorderly events. With the aid of an enormous bribe, he and his baritone costar, Antonio Scotti, were able to secure transportation out of the city.

Most of the citizens were stuck where they were. They milled about in the streets, wearing nightdresses or whatever they had thrown over themselves in their hurry to get out of their homes. Some reentered their homes to rescue the injured or to recover possessions. Some of these attempts proved fatal, as the almost continuous aftershocks brought down structures that had withstood the initial shaking.

The heavy masonry walls of City Hall collapsed, leaving the cupola perched incongruously on an iron latticework. Many other masonry buildings were destroyed. Still, the new skyscrapers along Market Street rode out the earthquake with little damage, and most wood-frame buildings—the vast majority of all the buildings in the city—survived, even if their chimneys cracked or fell. Only in the landfill areas near the waterfront did wooden buildings collapse in appreciable numbers.

The Scourge of Fire

But if wood construction gave protection against the earthquake's direct effects, it was all too susceptible to a related scourge—fire. Within half an hour of the earthquake, at least 52 separate fires had broken out around the city. Other fires started later as householders rashly lit their hearths under damaged or missing chimneys. Soon, two major conflagrations were under way, one north, one south of Market Street. And there was no water.

In this perilous moment, the Commanding General of the Department of California, Frederick Funston, seized power. He rounded up several thousand soldiers and had a proclamation printed (by hand, since there was no electricity) and distributed; it threatened summary execution for looting. Funston's men did in fact shoot or bayonet many looters, when they themselves were not looting. Eventually, Funston's authority was given the appearance of legitimacy by the three San Francisco newspapers. They put out a joint extra edition, which was printed across the Bay in Oakland; it stated (untruthfully) that President Theodore Roosevelt had proclaimed martial law in the city.

Funston's main effort was aimed at dynamiting buildings in the path of the fires, with the hope of creating firebreaks wide enough to stop the passage of the flames. This effort failed repeatedly on the first day and on the following day, too. It failed

THE QUAKE'S SCIENTIFIC SIGNIFICANCE

Not only was this one of the greatest earthquakes ever to hit California, it also was the most significant in many respects. It provided many clues to the nature of faulting and earthquakes in general, and the damage of man-made structures provided the basis for many building standards and rating tables still in use today. . . .

Even though the San Andreas Fault had been recognized as early as 1893, its size and importance were not really understood until the 1906 earthquake. But the extent of the movements at this time awakened international geologic interest in this great rift zone. For any single fault movement to produce a horizontal movement of 20 feet was unheard of at the time, and many scientists thought that the San Andreas Fault was unique in the world. Today, it is well known that most of the master faults around the circum-Pacific because the buildings selected for demolition were too close to the oncoming fire and because the dynamite ran out, leaving the soldiers nothing but gunpowder, which set fire to buildings more readily than it collapsed them.

Block after block was consumed. Mechanics Hall, a huge warehouselike structure near City Hall that had become an impromptu hospital, morgue, and repository for art objects, burned to the ground. So did the Palace Hotel, despite the three-quarters of a million gallons of water stored on the premises for just such an emergency. The skyscrapers on Market Street, the tenements of Chinatown, the wharves—everything that could burn, did so. Only portions of Telegraph Hill and Russian Hill were spared.

The Final Line of Resistance

When it became clear that the downtown area would be a total loss, the broad carriageway of Van Ness Avenue became the final line of resistance. Surely the fire could be stopped there and

---

basin are capable of similar horizontal shifts.

After the earthquake, a State Earthquake Investigation Commission, headed by geologist Andrew Lawson of the University of California, was appointed to study the earthquake. Its report, published by the Carnegie Institution of Washington, is the greatest report of its kind ever published on an earthquake. Its pages are still referred to by any geologist who wants to study not only the 1906 earthquake, but faulting in general. Its deep reservoir of facts has provided the research materials for countless articles and books.

This exhaustive study of the earthquake is generally considered to be one of the greatest contributions ever made to modern seismology. It set a pattern of investigation and reporting that has been followed ever since. It showed up the inadequacies of former attempts to study earthquakes, and pointed to the need for new research programs in earthquake causes and effects.

Robert Iacopi, *Earthquake Country*. Menlo Park, CA: Lane Books, 1976.

The earthquake of 1906 devastated San Francisco.

the residential areas to the west saved. Backfires were started in the elegant mansions along the east side of the avenue, but flames leaped across to the west side and ignited buildings there, too. It seemed as if the entire Western Addition was doomed, but thanks to heroic efforts by the firefighters, as well as by engineers who had partially restored the water supply, the fire was halted a block farther west, at Franklin Street.

The fire had raged for three days. A total of 28,188 buildings were destroyed in an area of 4.1 square miles. According to the Army's body count, 498 persons died in the earthquake or the fire; about four times as many were unaccounted for and probably also died. Property losses were $450 million in 1906 dollars (about $7 billion in current dollars). Recovery was slow and painful. And Caruso never did sing in San Francisco again.

Other Cities Devastated

The San Francisco earthquake devastated many other cities besides San Francisco. The City Hall of Santa Rosa, 50 miles to the north, was damaged even more severely than San Francisco's City Hall. Its domed tower fell onto the main building, crushing it completely. The recently completed buildings of Stanford University, 20 miles to the south of San Francisco, took terrible punishment; many of them were reduced to rubble. Buildings were damaged as far north as Humboldt Bay, near the Oregon border, and as far south as King City in Monterey County. The earthquake was felt over an extent of 175,000 square miles, from central Oregon to Los Angeles and from the coast to Winnemucca, Nevada. Even sailors at sea felt the shocks: some captains believed they had run aground, even though they were in deep water. The total number of deaths will never be known with certainty, but several thousand people must have perished.

The 1964 Good Friday Alaskan Earthquake

BY BRUCE A. BOLT

Alaska is the most seismically active area in the world. Situated along the northernmost arc of the Ring of Fire, Alaska sits on a subduction zone where the Pacific plate slides under the North American plate. The 1964 Good Friday Earthquake of Alaska, with a magnitude of 9.2, was the largest in United States history, and the second largest on world record. About 130 persons were killed and Alaska sustained $300 million in property damage. Rock slides, landslides, and snow avalanches damaged roads, bridges, railroad tracks, power facilities, and harbor structures.

In this article, seismologist Bruce A. Bolt outlines the activity of the earthquake, which began on the evening of March 27, 1964, under northern Prince William Sound and ruptured rocks along the Aleutian trench for eight hundred kilometers. He states that it was the greatest area of vertical displacement ever measured in earthquake history. The earthquake also caused large tsunamis, or seismic sea waves, that devastated waterfront developments along the Alaskan coast. The shaking earth changed soil and sand into a liquid state and carried away many modern frame houses.

T
he Aleutian Islands and trench stretch in a sweeping arc across the northernmost Pacific Ocean between Kamchatka in Siberia and south central Alaska. Into this trench the Pacific plate plunges downward and northward. Abundant volcanic and seismic activity occurs along the entire arc and extends eastward into the active and dormant volcanoes

of the Rango Mountains. Intermittent thrusting of the plate under Alaska occurs frequently, producing earthquakes over a wide region. The underthrusting plate may stick at any one place for centuries, while adjacent parts of it continue to progress onward. Finally, a break occurs.

The Earthquake Strikes

Such an event occurred on Good Friday, March 27, 1964, at 5:36 P.M. The first slip occurred at a depth of about 30 kilometers under northern Prince William Sound, and the rupture in the rocks extended horizontally for 800 kilometers, roughly parallel to the Aleutian trench.

Hundreds of measurements along the shoreline later showed that beds of barnacles and other sea life had been raised above sea level about 10 meters. From such observations and the uplift of tidal bench marks relative to sea level and from geodetic level lines surveyed carefully from the coast into Alaska, it was estimated that about 200,000 square kilometers of the crust were deformed in the Good Friday earthquake. It was the greatest area of vertical displacement ever measured in earthquake history. The subduction-zone slip occurred mainly beneath the ocean; only in a few places, such as on Montague Island in Prince William Sound, were fresh fault scraps visible. The vertical fault displacements on Montague Island amounted to 6 meters in places. Such great slips are not the record for Alaska, however; 14.3 meters of uplift occurred in the Yakutat Bay earthquake of 1899, centered about 320 kilometers to the east.

Gigantic "Tidal" Waves

The sudden upward movement of the Alaskan seafloor along the rupturing fault generated large water waves, acting on the water of the ocean like a gigantic paddle. Such gigantic "tidal" waves, produced in an earthquake, are called *tsunami*. The crests of the first waves struck the shores of Kenai Peninsula within 19 minutes and Kodiak Island within 34 minutes after the start of the earthquake. As the tsunami rushed onshore, it devastated waterfront developments along the Alaskan coast, particularly at Valdez and Seward. About 120 persons drowned.

In Anchorage, 100 kilometers from the fault slip in Prince William Sound, strong ground shaking commenced about 15 seconds after the rocks first broke. The heavy shaking continued for more than half a minute. After shaking began, the announcer at radio station KHAR, R. Pate, recorded his thoughts on a tape recorder:

A Radio Announcer's Thoughts

Hey, boy—Oh-wee, that's a good one! Hey—boy oh boy oh boy! Man, that's an earthquake! Hey, that's an earthquake for sure!—Wheeeee! Boy oh boy—this is something you'd read—doesn't come up very often up here, but I'm going through it right now! Man—everything's moving—you know, all that stuff in all the cabinets have come up loose. . . . Whooeee! Scared the hell out of me, man! Oh boy, I wish this house would quit shaking! That damn bird cage—oooo—oh man! I've never lived through anything like this before in my life! And it hasn't even shown signs of stopping yet, either—ooooeeee—the whole place is shaking—like someone was holding—Hold it, I'd better put the television on the floor. Just a minute—Boy! Let me tell you that sure scared the hell out of me, and it's still shaking, I'm telling you! I wonder if I should get outside? Oh boy! Man, I'm telling you that's the worst thing I've ever lived through! I wonder if that's the last one of 'em? Oh man! Oh—Oh boy, I'm telling you that's something I hope I don't go through very often. Maa-uhn!—I'm not fakin' a bit of this—I'm telling you, the whole place just moved like somebody had taken it by the nape of the neck and was shaking it. Everything's moving around here!—I wonder if the HAR radio tower is still standing up. Man! You sure can't hear it, but I wonder what they have to say on the air about it? The radio fell back here—but I don't think it killed it—Oh! I'm shaking like a leaf—I don't think it hurt it. Man, that could very easily have knocked the tower down—I don't get anything on the air—from any of the stations—I can't even think! I wonder what it did to the tower. We may have lost the tower. I'll see if any of the stations come on—No, none of them do. I assume the radio is okay—Boy! The place is still moving! You couldn't even stand up when that thing was going like that—I was falling all over the place

here. I turned this thing on and started talking just after the thing started, and man! I'm telling you, this
house was shaking like a leaf! The picture frames—all
the doors were opened—the dishes were falling out of
the cabinets—and it's still swaying back and forth—I've
got to go through and make a check to make sure that
none of the water lines are ruptured or anything. Man,
I hope I don't live through one of those things again. . . .

Damage Done

Building damage in southern Alaska from the 1964 earthquake
varied considerably, depending on the foundation conditions and
the type of structure. In Anchorage, higher buildings suffered
most, whereas frame homes were reasonably unscathed, although
their occupants were disturbed and furniture was thrown down.
Because of the distance from the rupturing fault, ground shaking
consisted mainly of long waves that do not affect small buildings.

In all, Alaska sustained 300 million dollars in property damage
from the earthquake; about 130 persons died, only 9 from the effects of shaking. One serious secondary result of the shaking was
the temporary change of soil and sand in many areas from a solid
to a liquid state. The most spectacular example of such *liquefaction* was at Turnagain Heights in Anchorage, where soft clay bluffs
about 22 meters high collapsed during the strong ground motion, carrying away many modern frame homes in a slide that re-
gressed inland 300 meters along 2800 meters of coastline.
Throughout southern Alaska, rock slides, land slides, and snow
avalanches were common, damaging roads, bridges, railroad
tracks, power facilities, and harbor and dock structures.

China's Killer Quake

BY RAYMOND CARROLL WITH HOLGER JENSEN,
SYDNEY LIU, LEONARD PRATT, AND TESSA NAMUTH

*The greatest killer earthquake of the twentieth century and one of the
most destructive in history, measuring 8.2 on the Richter scale, occurred in
Tangshan, China, in July 1976. Raymond Carroll and his associates
wrote the following article one week after the quake struck. In it, they de-
scribe the completely destroyed city of Tangshan, a coal and steel center of
more than 1 million people. The death toll was given at 200,000 people,
but later reports put it as high as 650,000 to 750,000. In addition, an
estimated 800,000 were injured and the regional economy was devastated. The Chinese seemed to accept the disaster with a stoic calm. Chinese scientists claimed to have successfully predicted a large quake the previous year, but this time they gave no warning.*

Raymond Carroll was a former editor of Newsweek. *Holger Jensen
is the international editor for the* Rocky Mountain News. *Sydney
Liu is* Newsweek's *Hong Kong correspondent. Leonard Pratt works as
a Chinese translator, and Tessa Namuth is a writer for* Newsweek.

Although few Westerners had even heard of the place,
Tangshan was a coal and steel city of more than 1 million inhabitants. Mining elevators rose out of the ground
in the middle of town, and far below the city streets men ham-
mered at the coal seams around the clock. But for all practical
purposes, Tangshan is gone. One night last week, as most of the
city slept, the world's worst earthquake in a dozen years tore it
apart. Hardly a building was left standing, and many thousands of
people almost certainly died. "The city was ruined totally, 100
per cent," said a shaken French traveller who somehow survived
Tangshan's obliteration.

Major tremors extended across populous Hopei Province in northeastern China and beyond. In Tientsin, China's third most populous city (4.3 million people), and in the capital of Peking (7.6 million), the earth shuddered and buildings cracked and crumpled. As people fled their homes and huddled in the streets under a torrential rain, they were shaken by [an] aftershock nearly as strong as the initial quake. By the end of the week, the authorities were warning of yet another powerful tremor, and millions of Chinese settled down in the streets as if for a long siege. Western seismologists measured the initial shock at 8.2 on the Richter scale—the strongest earthquake anywhere in the world since one measuring 8.5 devasted Anchorage, Alaska, in 1964. The second tremor measured 7.9 and must have finished off many buildings weakened by the first. The Chinese Government released no casualties figures, although the Central Committee of the Communist Party admitted that the quake had "caused great loss to people's lives and property." Foreign diplomats in Peking estimated the number of dead at 100,000 or more, and an American businessman who talked by telephone with a Peking trade official was given an estimate of 200,000 deaths. Whatever the figure, Tangshan's earthquake was far more lethal than the shocks that have shattered other parts of the world this year. Last February's quake in Guatemala registered 7.5 on the Richter scale and killed 23,000 people, while northeastern Italy's tremor of 6.5 caused 968 deaths in May.

No Warning

Earthquakes are an ancient enemy in China, but this one occured at a particularly bad time. It presumably crippled China's largest single source of coal at the outset of an ambitious industrial development program. It put further strain on a government that already has been weakened by the failing health of 82-year-old Chairman Mao Tse-tung and by intense factional disputes among his potential successors. The disaster also constituted an embarrassment for Chinese science. China has developed the world's most elaborate system of earthquake prediction, and in February 1975 thousands of lives were saved by the accurate forecast of a major earthquake near Yingkow in Liaoning Province. This time, there was apparently no warning.

Survivors of the Tangshan devastation said the quake struck suddenly and with terrifying force. The hotel in which a French friendship delegation was staying immediately collapsed in a heap of rubble. One 30-year-old Frenchwoman was crushed to death when a concrete wall caved in. Miraculously, the 22 other members of the French group managed to pick their way through the debris to safety in their bare feet. Later, in Peking, where they were flown by special plane, some of the Frenchmen still seemed to be in a state of shock. "It was horrible," recalled 60-year-old Maurice Monge. "We were lost. It was like an ocean, an ocean, everything moving."

Nine Japanese technicians of Hitachi Ltd., also were asleep in a hotel when the earthquake hit. Three of them were killed, and one of the survivors, Kinya Toyama, tried to crawl under his bed and was buried under the rubble. "I was suffocating and could hardly breathe," Toyama later related. "Then another jolt tossed the bed and debris into the air and I managed to crawl out." The French and Japanese agreed that few of Tangshan's buildings remained standing, and Western diplomats in Peking were convinced that few of the city's 1 million inhabitants could have escaped death or injury. Many thousands may have been lost in the coal mines alone, where the night shift was hard at work undergound when the earthquake ripped into Tangshan.

Terrifying

The port city of Tientsin, about 60 miles away, also suffered extensive damage. Former Australian Prime Minister Gough Whitlam and his wife—in the city at the end of a worldwide sightseeing tour—fled the nine-story Tientsin Friendship Guest House in their pajamas. "It was absolutely terrifying," Mrs. Whitlam said later in Tokyo. "The building was grinding and bucking and we were crawling around in the darkness." When Mrs. Whitlam's leg was cut by a glass from a shattered mirror, her husband bound it up with a towel. Then the two visitors followed a Chinese official with a flashflight down the broken stairs. "Outside, people were digging under the rubble," said Mrs. Whitlam. "Whole facades of buildings had come down." The guest house itself, her husband reported, was "literally split down the middle with a 1-foot gap separating the two parts."

In Peking, 100 miles from the quake's epicenter, older buildings crumbled and large cracks appeared in the facade of the Chinese capital's main department store. But most of the residents of Peking were merely shaken and showered with falling plaster and broken window glass. Only about 50 people in the capital were killed by the initial shock, and in the days that followed much of the city's population moved out of their houses onto the treelined streets. For a time, torrential rains fell on Peking, and the city soon took on the appearance of a giant, drenched refugee camp. Stoves, chairs and beds were dragged out into the street and people crouched under makeshift canvas shelters and gaily colored umbrellas, washing, cooking and eating as though they were still at home.

Ordinary Chinese seemed to accept the disaster with stoic calm. In Tientsin, seventeen American tourists were led in the dead of night from their damaged hotel to the safety of the city square. As dawn broke, the Americans were astonished to find themselves surrounded by thousands of Chinese, who had been sitting there in utter silence. Later, when their train to Peking stopped in a small village, the Americans were startled by another tremor. An old man on the station platform gestured soothingly to them with his hands, urging them to stay calm.

Taxis and Pajamas

Some Peking residents, including members of the foreign community, returned to their homes after the first two shocks. But the government was taking no further chances. Citing "unusual phenomena and continued tremors," officials gave warnings of a possible third quake. Men blowing whistles, banging kettles and shouting "Earthquake, earthquake!" aroused foreigners from their sleep and told them to quit their apartments—or leave the city altogether—at once. The evacuation was handled in an orderly manner. Foreigners were assigned to numbered taxis and were driven to their respective embassy compounds, where they camped outdoors. Some British diplomats spent one night sitting in their pajamas in the middle of a tennis court.

So far, [1976's] earthquake activity around the world does not appear to be unusual, although there may have been a larger than average number of victims, some experts say, because the quakes have struck more populous areas. "Routinely, we locate 5,000 to 6,000 quakes a year worldwide between magnitudes of 2 and 8," said Carl Stover of the U.S. Geological Survey in Golden, Colo. "But many smaller ones occur each year, probably 50,000."

To spot impending quakes, the Chinese use thousands of "peasant observers," as well as hordes of highly trained scientists.

Using folk wisdom passed down through 3,000 years of earth-quake watching, the rural "barefoot seismologists" look for nat-ural signs: changing water levels in wells, horses refusing to eat or enter the barn, birds reluctant to land, cockroaches twirling about in crazed indecision. The scientists monitor such esoteric data as minute movements across geological fault lines, slight changes in the tilt and electrical resistance of the earth's crust and decreases in the amount of radioactive gas in the well water. This infor-mation and more is gathered at 5,000 separate locations across China and fed into seventeen fully equipped seismographic cen-ters and 250 auxiliary stations.

Chinese scientists claimed to have saved thousands of lives by their predictions, and U.S. geophysicists who have visited China in recent months are convinced that a major disaster was indeed averted in Liaoning last year, when residents were sent outdoors before the quake struck. But the American scientists also point out that there were several warning foreshocks before that tremor. "We think that is the key to their prediction," said Jim Savage of the Geological Survey in Menlo Park, Calif. The Chinese did make a general forecast that there would be a major quake in the Peking-Tientsin area before 1980. But this time there were no foreshocks, and China's scientists issued no alarm.

Nonetheless, Chinese authorities reacted with speed and effi-ciency. Foreigners airlifted from the seaside resort of Pehtaiho, 70 miles northeast of Tangshan, said they saw long columns of trucks headed for the stricken city loaded with prefabricated houses, temporary bridges and shovels to dig out the dead and dying. Doctors and nurses were quickly dispatched to the Tang-shan area, while earthquake victims were rushed from the de-molished city to Peking's hospitals. Communist Party headquar-ters in Peking urged the people and the army to "plunge into anti-quake fight with a firm and indomitable will . . . guided by Chairman Mao's revolutionary line."

Stricken Area

Despite such invocations, China may find it difficult to recover quickly from the Tangshan earthquake. The tremor presumably damaged the vital oil pipeline from Manchuria to Tientsin and Peking. Tangshan's mines, which produce 6 per cent of the na-tion's coal, may be out of operation for some time. The stricken area is also an important industrial basin in a country whose industry has little resilience, and the quake could be a sizable set-back for China's new Five-Year-Plan.

The giant tremors also could reverberate on a psychic level, since many of China's people still attach great significance to natural omens. In the past two years, China has suffered several large earthquakes and meteor showers. After those phenomena, the revered Premier Chou En-lai died last January and his heir-apparent, former Deputy Premier Teng Hsiao-ping, was suddenly overthrown, reflecting a bitter power struggle in ruling circles. The new quake may be interpreted widely as a sign of even worse trouble to come. Even in the reborn China of Mao Tse-tung, a quake as immense as the one in Tangshan has a specific meaning to the superstitious. It foreshadows, they believe, the impending death of an emperor.

Quake-Ravaged Turkey

BY RICK GORE

The North Anatolian Fault is a rip in the earth's crust that runs a thousand miles from eastern Turkey to Greece. It is comparable to California's San Andreas Fault in that it has two large tectonic plates rubbing against each other—the Eurasian plate and the smaller Anatolian plate, which carries most of Turkey. On August 17, 1999, a devastating earthquake struck along this fault. National Geographic *senior assistant editor Rick Gore visited Gölcük, a Turkish city near the quake's epicenter, and İzmit, the largest city hit by the earthquake. In this selection, Gore reports that the infrastructure of one of the most industrialized regions of the country was completely destroyed and that the tragedy touched almost everyone in the country. The official death count was just over 17,000 people but actual fatalities may have been twice that number. In addition, 250,000 people were left homeless and 85,000 buildings were destroyed or uninhabitable. Personal accounts emphasized the human tragedy of the event and disgust of Turkish people concerning their government's shoddy building practices, which were responsible for much of the death and destruction. Citizens, however, were encouraged by the outpouring of aid from around the world.*

Three months later, the North Anatolian Fault struck again in a much less populated area but still killed more than eight hundred people. Sections of the fault are still expected to break and Istanbul, a city of more than 7 million people, is only a few miles from a possible epicenter.

The night was too sultry for sleep. So at 3 A.M. on August 17, 1999, restless townspeople in Gölcük, Turkey, strolled in the park. They walked along the waterfront on the Gulf of İzmit, easternmost arm of the Sea of Marmara, and perhaps some talked of the many jumping fish spotted on the coast

Rick Gore, "Wrath of the Gods," *National Geographic*, vol. 198, July 2000, pp. 32–52.

in recent days—or of the mysterious appearance of dead crabs and jellyfish. But İzmet Koyun and six other members of a local soccer team reminisced about past games as they sat on benches beneath a willow tree.

"Let's go home," İzmet recalls saying. "I've got to work today." As he stood, an explosion boomed over the gulf. "The earth came alive with shaking," he says. "The sky turned red, a sword of light flew out of the sea, and a wave as tall as a ship thundered toward us."

A great crack opened along the waterfront, and "like a drunk man trying to run," İzmet leaped over it and raced inland. Three of his friends climbed trees. A blinding storm of dust from collapsed buildings rose over the town and swept down toward the shore.

When the dust settled, İsmet found himself knee-deep in water. Looking back, he saw that a vast section of the former waterfront, including the park, had slumped into the gulf, sinking 30 feet or more. The lower floors of two seven-story buildings had crumbled and plunged into the gulf, killing 50 men who had been gambling in a ground-floor café.

Like everyone I met along the Sea of Marmara in the days after the earthquake, İzmet Koyun still seemed stunned as he sat on his bicycle and looked out at the submerged trees and lampposts now well offshore. "Many bodies remain out there under the water," he said. "Also many vehicles, including a police car with two policemen inside."

Gölcük lies near the epicenter of one of the most punishing earthquakes of the past century. The magnitude 7.4 catastrophe created headlines worldwide. Tens of thousands dead. Some 250,000 homeless. And billions of dollars' worth of damage to Turkey's industrial heartland.

Istanbul, a city of more than seven million people about 50 miles northwest of the epicenter, was violently shaken. Although the heart of the megalopolis remained intact, the quake destroyed several dozen buildings in Avcılar, a neighborhood built in recent decades on the western edge of the city. And thousands of people, too frightened after the quake to sleep indoors, camped in open spaces with tents thrown together from sheets, towels, and blankets.

"This tragedy has directly touched almost everyone in the country," said my Turkish friend Aydın Kudu. "Hundreds of thousands of people from all over Turkey had moved here for jobs. We've all lost someone."

Where the Earthquake Began

The earthquake began just east of Gölcük, about ten miles underground along a buried rip in the Earth's crust known as the North Anatolian Fault. Extending from eastern Turkey to Greece, the thousand-mile-long rip is very similar to California's infamous San Andreas Fault. Like its American counterpart, the Anatolian Fault is actually a network of smaller fault segments that divide two tectonic plates—in this case Eurasia and the much smaller Anatolian block, which carries most of Turkey on its back.

The edges of the two plates are locked together, but geologic forces are driving the Anatolia plate westward toward Greece at the rate of about eight or nine feet a century, building pressure along the juncture. When enough pressure builds, one or more fault segments unlock in a violent jerk. If a small segment breaks, the ensuing earthquake might be magnitude 6 or less. But when the segment beneath Gölcük snapped, the energy released triggered ruptures along three adjacent segments—to both the east and west—creating a much larger event.

Three days after the earthquake struck, I drove with Aydin toward the epicenter, reading the morning headlines. The official death count had reached 6,800. Three provincial governors—unable to coordinate initial relief efforts—had been replaced. Rescue workers, including at least 2,000 foreigners, were giving up hope of finding anyone else alive in the wreckage. In the city of Adapazarı 963 people had been buried in a mass grave. The Turkish government was ordering 10,000 more body bags. The destruction of the infrastructure of one of the most industrialized regions of the country was "complete," said the general secretary of the Foreign Investors Association.

Damage in İzmit

When we reached the edge of İzmit, the largest city hard hit by the earthquake, the smell of petroleum pervaded the air. Black smoke still billowed up from the Tüpras refinery complex, Turkey's largest. People walked the highway with suitcases. Relief workers carried corpses to an ice rink that had been converted into a morgue.

We came to an immense pile of broken concrete—the remains of a six-story apartment building. Carpeting, bedspreads, and splinters of furniture protruded from the rubble. A rescue team working with a large backhoe picked away at the debris pile.

Clustered around the collapsed building were scores of people, their eyes red, their faces weary. Many clutched photo albums or stared at pictures of loved ones, hoping against all odds that they might still be breathing beneath the concrete.

A shout arose from the rescue team—a body. The workers carefully extracted the corpse of a woman; people crowded around. No one recognized her.

"It's hard," said a bystander. "The faces are so swollen and black."

I felt intrusive and helpless. "What can I say to these people?" I asked Aydin. *"Geçmis olsun,"* he replied. "May it be over."

An older man with bandaged head and arms arrived, and people rushed to greet him. He was Mustafa Çifttepe, one of seven people who had survived the collapse of the building. His wife and son had not. He was just back from the hospital.

"He was trapped for 17 hours," said his son-in-law, Ersin Güzey, who lives in New York City. As soon as they heard of the catastrophe, Ersin and his wife had caught the first flight available to Istanbul.

Ersin translated his father-in-law's story: "I was in bed with my wife on the second floor when the shaking began. My wife said: 'Get up! Get up!' Then a large chest landed on her. A wall fell toward me, stopping two inches from my nose. I was afraid to move, afraid if I did, everything would collapse and crush me. I just prayed and called for my son."

His son, a neighbor named Vedat Aktas told me, was found crushed in the next room, his pants pulled halfway up his legs.

Vedat cursed the contractor who built the apartment. "This building had only half the steel it should have had," he said. "And this is supposed to be concrete," he said, dropping a chunk and watching with disgust as it shattered into bits. "It's more like sand."

Vedat's anger echoed across Turkey as the death toll mounted. The dire need for housing during recent rapid urbanization had encouraged fast, shoddy construction in thousands of new buildings. Many contractors, either through corruption or negligence, ignored building codes designed for earthquake resistance. It was those newer buildings that took the greatest hit.

As we left, I turned to the bandaged old man: "Geçmis olsun." He nodded, his eyes filling with tears.

Deeper into the Destruction Zone

We headed deeper into the destruction zone, giving a ride to a man bound for Gölcük. His niece and her husband had died there, and he wanted to search for other missing relatives. The sun blazed. The air sweltered. And everywhere I looked I saw the heaped remains of countless shattered stores, mosques, and apartments. Most of the buildings still standing teetered on the brink of collapse. Traffic inched past bulldozers working to clear debris. Thousands of sweating, overworked rescue workers attacked concrete rubble with jackhammers. The smell of death filled the air, and, like most of the people we saw, we put on surgical masks to filter the dust, odors, and microorganisms.

Aydin and I stopped to let our hitchhiker out and met two schoolteachers—a husband and wife—sitting in a field surrounded by an upholstered couch, a dining room set, and various other pieces of furniture they had retrieved from a nearby apartment building. They pointed to where they had lived.

"We were on the fifth floor," said Gönül Güzel. "Now it's the third. The top floors fell into the first two. Everyone on the bottom died."

"We have no idea what will happen to us," said his wife, Yücel. "People bring us food, but we have no desire to eat."

At the resort town of Yalova we met a friend of Aydin's named Hakkı Akyazı. His eyes were swollen, and in a soft, slow voice he told us he had just buried his sister, a medical doctor who had lived in a new apartment complex there. The two had been on vacation together on the Aegean. She had decided to extend her stay and had returned to Yalova for the night to pick up more clothes.

When Hakkı heard of the earthquake, he rushed to her building, which lay in ruins. No rescue teams had reached it. He figured out where her apartment would have been and for 12 hours worked alone to get inside her bedroom. When he finally did, he found her dead.

At dusk we took a ferry from Yalova back to Istanbul across the Sea of Marmara. As we left the dock, I watched dump trucks pull up to the shore and empty the rubble of so many lost lives into the water. There was no other place to put it. Our boat was packed with mourners, and the setting sun turned the water deep red. It seemed as if we had been to a thousand funerals all in one day.

Speaking with Geologists

The next morning we visited Istanbul Technical University to speak with geologists struggling to understand exactly what had happened along the North Anatolian Fault.

"We knew that the Gölcük area was where the fault was likely to break next," said Rob Reilinger, an American geophysicist from MIT. Reilinger uses the satellite-based global positioning system (GPS) to track deformations or swellings of the Earth's surface that indicate where pressure is building underneath. "We just didn't know when this earthquake would happen or how much of the fault would break."

Scientists are able to explain some earthquake phenomena that puzzled the people of Gölcük. The flash of light that İsmet Koyun saw over the Sea of Marmara may have been methane gas exploding as it was released from sediments in the gulf. The dead crabs and jellyfish seen at Gölcük could have been killed by radon gas seeping from rock in the Earth's crust into the water just prior to the rupture.

But there is much the scientists still do not understand, and much that disturbs them. This earthquake, explained geologist Aykut Barka, was part of a sequence that began along the North Anatolian Fault in 1939 near its eastern end. Historically each section of the fault in Turkey breaks on average every several hundred years. Before this earthquake the only stretch of the main fault that had not broken in the 20th century extended from the city of Bolu, about 90 miles east of Gölcük, to the western end of the Sea of Marmara—a distance of about 220 miles. The four fault segments that broke on August 17 account for only about 70 miles of the 220 at risk. This earthquake probably pumped additional strain into unbroken segments.

Threat to Istanbul

Most worrisome are about a hundred unbroken miles of the fault that lie deep beneath the Sea of Marmara, passing less than 15 miles from Istanbul. Scientists do not understand well the structure of that obscured stretch, but they know it poses dangers. Nicholas Ambraseys, a specialist in historical earthquakes at Imperial College in London, notes that 40 earthquakes of magnitude 7 and above have hit the Marmara region since the first century A.D. In 1509 and again in 1766, great earthquakes destroyed much of Istanbul. Both may have been part of a 250-year rupture cycle. Some experts now argue that one or more events at least as large as the August quake will occur in the sea south of Istanbul in coming decades.

How bad might the next Istanbul quake be? That depends in part on how far the epicenter is from the city. It also depends on whether the fault segments beneath the Sea of Marmara break together or independently. Together they could create an earthquake as strong as magnitude 7.8—about four times stronger than the August earthquake.

Some of the Worst Damage

A week after the earthquake I received permission from the Turkish Navy to walk where the North Anatolian Fault did some of its most spectacular damage: the Gölcük Naval Command Center, the largest naval base in Turkey. When the quake tore through the Gulf of İzmit, it devastated the compound, toppling buildings and killing hundreds of people.

"I've been in many earthquakes, but nothing like this," recalled Ercüment Doğukanoğlu, a naval captain. "When it hit, I felt helpless—like being thrown every which way in a frying pan."

Heavy rain fell as a young second lieutenant, Selçuk Poyraz, led us across the ravaged base. "The rain is nice," said Selçuk. "It washes away the smell of death, which gets into everything. I have to put cologne in my car, on my clothes."

He walked us to a green lawn and pointed at what looked like the burrow of a gigantic mole cutting across the base, creating a scarp several feet high. In its path lay the ruins of an officers' club, where scores died in their sleep.

We followed the burrow until it opened into a crack so wide I had to jump to cross it. We reached a place where the crack had split a stone wall, thrusting the south end eight feet to the west into the middle of the street it once bordered. The same crack also pushed an entire apartment building across the street several feet closer to Greece.

As the earthquake faded from the world's headlines, its miseries persisted. I returned to Turkey in late October. The official death count stood at just over 17,000, but the real toll may have been twice that. Winter was setting in, and although people were being fed adequately by organizations such as the Red Crescent, warm clothes remained in government warehouses undistributed. With more than 85,000 buildings destroyed or uninhabitable, about 40,000 families were living in 168 tent cities. Few tents were winterized.

Psychological problems were mounting. Many men remained jobless and idle. Many of the neighborhood coffeehouses they relied on for socializing were gone, replaced by scattered coffee tents. Children played and attended school in severely over-crowded tents. So many teachers died in Adapazarı that in one district only two remained to look after 2,000 children.

As a group, the women displaced from their homes may have been suffering the most. "They have no place to go to be together," said Mebuse Tekay, a relief coordinator for 128 non-government volunteer groups.

Some Positive Changes

Despite the crisis, Mebuse pointed to positive changes the earth-quake brought.

"It has collapsed some taboos," she said. "Many Turkish people thought they had no friends in the world except other Turks. But so many foreigners came to help us, we now must see a new re-ality. Even the Greeks proved not to be our enemies. Television showed a Greek team crying after they rescued a little girl.

"Also many of the people rushing to help were from arts groups. Now many traditional Turks have had to change their bi-ases toward men who have long hair or earrings and women in miniskirts. It was those people who showed up first."

The Fault Strikes Again

On November 12 the North Anatolian Fault struck again. Stress from the August quake triggered a rupture along a segment of the fault east of the earlier break. The second quake measured magnitude 7.1. It hit a much less populated region but still killed more than 800 people and injured at least 5,000.

In the town of Kaynaslı, Özgür Akbulut's father and older brother had just left evening prayers at the mosque when the temblor hit. As most of the town's buildings buckled, the mosque's towering minaret crashed down on the two men, crushing them to death.

On a cold and drizzling December morning I met 15-year-old Özgür in the nearby village of Handanoğlu, where he was living in a tent with surviving relatives. As I talked with the men of the village in the parlor of a farmhouse that had survived the quake, Özgür sat silently by the stove, warming his hands. He smiled occasionally but mostly stared vacantly ahead.

"These earthquakes are tests from God," said Mehmet Bayındır, a wizened 92-year-old. "We should build houses the old way—from chestnut wood. They don't collapse."

His grandson Hüseyin agreed. "I accept that this was a geologic event, but it can be taken as a warning. In seconds, billionaires can become penniless. So you must have values that you can't lose—a good heart and honesty."

Although smaller, Kaynaslı looked like Gölcük all over again—streets lined with shattered buildings and mournful people struggling to rebuild. I headed back to Gölcük, about 80 miles away, to see how people there had coped in the intervening months.

The nightmare hadn't gone away. Bulldozers had cleared most of the piles of rubble, and temporary prefabricated houses had risen rapidly outside the town. But Gölcük itself was still a city in shambles. I found İsmet Koyun again in a coffeehouse by the waterfront.

"There's not much else to do," he said as we drank tea with his friends. "Gölcük is dead. Most people have left. The government hasn't decided whether the city should be rebuilt."

At dusk, İsmet took me to the water's edge, where he had watched his city collapse. We clambered over twisted rebars and mud-caked chunks of concrete to the seven-story building where the 50 men had perished in the sunken café. We watched the water lap into what had been the building's third floor.

"Thirteen bodies were never found," he said. Darkness fell, and I could only think of one thing to say: "Geçmis olsun. May it be over."

Averting Disaster

The Difficulties of Earthquake Prediction

By Mark Buchanan

According to author and physicist Mark Buchanan, earthquake prediction has always been a challenge. Countless earthquakes have been predicted that never occurred, causing governmental panic, and large earthquakes have broken loose with no warning whatsoever. One reason for this difficulty may be a tendency for the earth's crust to "self-organise" into what is called a "critical state," a condition of instability in which it is poised on the edge of change. When the earth is in this state, it becomes impossible for scientists to know what will happen next. The implication that earthquakes cannot be accurately predicted suggests that scientists should work on different ways to calculate some general probabilities.

W hy can't scientists warn us about earthquakes such as the one that struck in Gujarat, western India [in January 2001], killing possibly as many as 100,000 people? Why is it—if astronomers can see nearly to the edge of the universe and biologists can clone living organisms—that the science of geophysics cannot tell us when and where the earth will start shaking?

What is most peculiar about this situation is that the basic earthquake process is conceptually simple. The continental plates are great fragments of the earth's crust which float on a liquid mantle like gigantic rafts. Wherever two of these rub shoulders—as they do all along the San Andreas fault in California, for instance—they tend to stick together. But slowly, as continents drift, the rocks get twisted out of shape and, when the stress builds up

Mark Buchanan, "Is Earthquake Prediction Just Literature?" *New Statesman*, vol. 130, February 2001, p. 16. Copyright © 2001 by *New Statesman*. Reproduced by permission.

beyond a certain threshold, something finally gives and there is an earthquake.

So, it seems, it shouldn't be too difficult to tell where and when the big events will take place. This isn't quantum physics. Even so, the tragedy in India reminds us that the record of earthquake prediction research is truly dismal. There has never been a single unambiguous success. And there have been many notable failures.

Some Prediction Failures

In 1976, for example, a researcher with the US Bureau of Mines predicted that two enormous quakes of magnitude 9.8 and 8.8 on the Richter scale would strike off the coast of Peru in August 1981 and May 1982. He also predicted a foreshock of magnitude 7.5 to 8 for June 1981. When the foreshock didn't happen, he retracted his prediction, but the Peruvian government was thrown into such a scare that an official of the US Geological Survey had to travel there to calm fears.

Also in the late 1970s, Japanese scientists became convinced that a great quake was soon to hit central Japan. In the past, earthquakes had occurred in the region at roughly 120-year intervals. As more than 120 years had passed since the last quake, they reasoned that another was imminent. A vast emergency response system was duly put in place and yet, today, 25 years later, the quake has still not arrived. The 1995 Kobe quake, in fact, struck in an area of Japan where scientists thought the risk was low.

In 1995, the chairman of the geology department at the University of Southern California predicted that a large quake would rock central California in the spring or early summer of that year. It never happened.

There are countless examples of similar failures. Robert Geller of Tokyo University, one of the world's foremost earthquake experts, wrote recently: "Earthquake prediction research has been conducted for over 100 years with no obvious successes. Claims of breakthroughs have failed to withstand scrutiny. Extensive searches have failed to find reliable precursors . . . reliable issuing of alarms of imminent large earthquakes appears to be effectively impossible."

Can an area of research even be considered to be scientific if it cannot make predictions? As the poet Paul Valery saw things, "'Science' means simply the aggregate of all the recipes that are always successful." And to that he added, "All the rest is literature." Does earthquake science belong to literature? Should earthquake researchers throw in the towel and work on something else? In an online debate hosted by the science journal *Nature* in 1999 (www.nature.com), one geophysicist suggested that earthquake prediction is "the alchemy of our times", a topic that, despite its practical impossibility, is "fatally attractive to both scientists and the general public".

The Unstable "Critical State"

Over the past decade, physicists have discovered that systems as diverse as a pile of sand, the earth's crust and its ecosystems, and even our financial markets seem to have a tendency to "self-organise" into what is known as a "critical state". This is a natural condition of extreme instability in which the system remains always poised on the edge of sudden, radical change. It is, in a way, tuned so as to be hypersensitive to even the tiniest of influences. In such a setting it becomes next to impossible to predict what will happen next.

The statistical distribution of earthquakes—the raw numbers for how many small, intermediate and large quakes take place—follows a mathematical pattern consistent with this idea. What's more, researchers have discovered that some of the basic models which geophysicists use to mimic the earthquake process also fall into this category.

The idea suggests that the pattern of stresses and strains in the earth's crust is not random, but possesses an intricate organisation which makes it extremely difficult to foresee the ultimate consequences of, say, a tiny increase in stress in just one place. That change might do nothing more than bend and compress the rocks a bit more. Or it might set off a small quake—a sequential slipping of rocks along a fault that would carry on for a short distance. It could, on the other hand, set off a chain reaction of movement going much further and resulting in a major quake.

If this idea is correct, then highly intricate features of the global organisation of stress and strain within the earth will determine where and when the next great quake will take place. This implies that there may be no identifiable details in the crust from which scientists could hope to "read out" the future, and that no amount of data gathering and computation will ever be enough to make a prediction even of the rough timing or magnitude of the next earthquake.

As one geophysicist put it, an earthquake when it starts out "does not know how big it will ultimately become". And if it doesn't know, neither can we.

This view suggests that it may be wise for geophysicists to work in a different way. Even if one cannot make predictions, this does not imply that there is no regularity at all in the earthquake process.

What Can Be Known

We know, for instance, that earthquakes cluster along those places where continental plates meet. This leads to the uninspired conclusion that it is more hazardous, as far as earthquakes are concerned, to live in San Francisco than in London. But quakes also cluster in time, and this teaches a more surprising lesson. It is a seemingly intuitive idea that if an area suffers no earthquake for a long time, then it is "overdue" for a quake and should suffer one soon. But this flies in the face of the actual statistics. On the contrary, the longer a region goes without a quake, the less likely it is to see one soon.

These ideas suggest that we may be able to benefit from something less ambitious—predicting not the placement and location of specific quakes, but the likelihood of having a quake of a certain size in a certain zone over a certain number of years. These aren't the kinds of predictions that grab headlines but they may be enough to inform building codes and to establish good procedures for emergency response in areas that are indeed subject to hazard.

Friedrich Nietzsche put his finger on the psychology that makes us long to find readily identifiable causes for catastrophes: "To trace something unknown back to something known is alleviating, soothing, gratifying, and gives moreover a feeling of power. Danger, disquiet, anxiety attend the unknown, and the first instinct is to eliminate these distressing states. First principle: any explanation is better than none . . . The cause-creating drive is thus conditioned and excited by the feeling of fear . . ."

As the tragedy in India has once again emphasised, however, there may be no simple answers; indeed, there is good evidence that predicting the circumstances of catastrophes may in many instances be strictly impossible. Unforeseen catastrophes are a grim feature of reality, and they may be with us always.

Improving Post-Earthquake Search and Rescue Operations

By Larry Collins

Los Angeles fire captain and rescue specialist Larry Collins empha-sizes in this selection the need to prepare for large seismic events in the United States. He refers specifically to California, Alaska, Washington State, the New Madrid Fault Zone in Missouri, and other places where the earth moves regularly. In regard to post-earthquake rescue operations, Collins stresses the importance of continuing the search for survivors in collapsed buildings for at least two weeks following the disaster. He cites instances of people and animals found two to three weeks later in "void spaces" honeycombed throughout collapsed buildings. Collins explains what further steps should be taken in search and rescue operations and that disorganized emergency response, failed communications, inexperi-enced rescuers, and misinformed officials are sources contributing to disas-trous situations worldwide.

One of the most important lessons from recent earth-quakes in Taiwan (September 1999), Athens (Septem-ber 1999), and Turkey (August 1999) is this: No single government has all the answers or all the resources to manage the worst earthquake disasters likely to strike in the future. Based on the damage sustained in these three earthquakes, as well as those in Kobe (1995), Armenia (1988), and other seismic hot spots, Southern Californians and others living in quakeprone regions should be prepared for something exceptionally horrific, including life loss that may greatly exceed any previous U.S. disaster.

Every damaging earthquake is a sort of library, a storehouse of lessons for those in search of them. The problem is quantifying these lessons in a meaningful way, then determining how society should weave them into the fabric of modern daily life, commerce, and governance in a manner that will prevent the unnecessary loss of lives and property in future seismic events.

It's a sad reality that many lessons from contemporary quakes are actually a rehash of those that should have been gleaned from past events. The continuing conflict between embracing and ignoring information from damaging earthquakes is an indication that modern society hasn't fully developed the ability to absorb the most important elements of these lessons.

The United States rode a figurative wave of seismic good luck through the 20th century. Despite a series of moderate quakes that caused serious damage and more than 100 fatalities and a Great Quake that killed hundreds (some experts say thousands) in San Francisco in 1906, the United States has yet to suffer a catastrophe on the scale of the recent earthquakes in Turkey and Taiwan, where tens of thousands died in collapsing buildings.

The 7.1 Hector Mine earthquake that struck Southern California on October 16, 1999, was indicative of the frail balance between good and bad seismic fortune in the United States. The quake originated on a fault once thought to be inactive, tearing through 25 miles of desert floor near Twenty Nine Palms, 100 miles east of Los Angeles, displacing the ground by nearly 15 feet. If a rupture on that scale had occurred beneath virtually any heavily populated spot in the Greater Los Angeles area, firefighters and urban search and rescue (USAR) teams might have been struggling to locate and extract live victims from the rubble of collapsed buildings two weeks later. But this was a sparsely populated desert, and not a single person died.

Recently, it was announced that a new blind thrust fault capable of a 7.3 earthquake has been discovered in Southern California's Orange County. It is one of many newly recognized earthquake faults capable of causing catastrophe.

Such discoveries should serve as reminders to intensify preparations for the major seismic events likely to strike the United States during the next century, including the potential for one or more truly large quakes that could cause devastation across thousands of square miles in places like California, Alaska, Washington State, and New Madrid Fault Zone.

The most recent quakes in Turkey caused several coastal towns to literally sink beneath the sea. The tsunamis that followed flooded several coastal zones. Similar effects are expected along the West Coast of the United States. The shaking collapsed thousands of buildings, killing at least 15,000 people. Two weeks later, across the Aegean Sea, Athens, Greece, was pounded by a quake that collapsed dozens of modern buildings. In Asia, entire cities have been leveled by quakes that killed untold numbers of people, the latest of which struck Taiwan. Modern Asian cities have been severely damaged by tremors whose ferocity architects and builders never anticipated. Similar conditions are found in places like Los Angeles, San Bernardino, San Francisco, and Seattle.

When I was dispatched as part of a team to observe Japanese emergency operations following the Kobe earthquake, the ease with which the shaking toppled and crushed modern "Western-style" steel and concrete buildings left a vivid impression. Another shock was witnessing the devastating effect of firestorms that swept large swaths of the city and burned people alive as they lay trapped in collapsed buildings.

Potential Impending Disasters?

The potential for seismic disaster is seen wherever the earth moves with regularity. In Chile and Mexico, great quake-spawned tsunamis swept entire coastal areas clean of human habitation in 1996. In Mexico City, which lost more than 10,000 residents to a 1985 quake, millions of people live in structures built on an ancient lake bed that sometimes accentuates the ground motion of earthquakes. In terms of seismic risk, it is a city built for disaster.

During the 1800s, the flow of the Mississippi River was for a time reversed by a series of earthquakes in the New Madrid Fault Zone—quakes of a size and ferocity that seldom have been witnessed. In those days the fault zone was sparsely populated, so the human effect was not particularly evident. But today, this region is densely clustered with apartments, high-rises, industrial complexes, and chemical factories erected by designers who apparently forgot or ignored the lessons of the New Madrid earthquakes. A similar event today would cause a catastrophe unprecedented in U.S. history.

In Los Angeles, the 1994 Northridge quake struck on a blind thrust fault that wasn't even known to seismologists. The eastern end of the Santa Susana Mountains grew more than a foot in a matter of seconds and rose several more inches during ensuing aftershocks. The thrusting fault alternately raised and lowered large chunks of the San Fernando Valley, permanently changing the elevation of many neighborhoods by as much as nine inches.

The landmark Palos Verdes peninsula in Southern California is actually a fold of the ocean floor: it grows ever higher as the crust is compressed several inches every year by the collision of the Pacific and North American plates. Twenty-six miles offshore from Los Angeles, Catalina Island and her offshore sisters are a product of great tectonic forces. Just to the east of Los Angeles, the towering San Gabriel Mountains owe their steep ascent to a locked "dog leg" section of the San Andreas Fault, which pushes the mountains upward much like one might squeeze folds of skin together on the back of one's hand. The 10,000-foot-high San Gabriels are growing faster than erosion can wear them down. Similar dynamics are in evidence across Southern California.

In some ways, the Greater Los Angeles area is one giant tectonic laboratory, where all possible effects of earthquakes—and man's societal response (including emergency services)—are tested in real time.

Search and Rescue Operations

The average person—and even some emergency responders—might assume that no one can survive in large buildings that have collapsed into piles of rubble after an earthquake. The subsequent rescue of trapped victims after the first day is almost universally a surprise and is characterized as "miraculous" by the media, politicians, and the public. Consequently, search and rescue operations often taper off after the first day. People simply don't expect survivors after Day 1. Yet, many collapsed buildings (the collapse might be caused by any phenomenon, not just earthquakes) are honeycombed with void spaces in which victims might be trapped alive. There is indisputable evidence to support this view.

In Mexico City, many people—including infants in a collapsed hospital—were rescued alive after more than a week of entrapment. Following the Armenia quake, survivors were rescued after nine days. In the Philippines quake, a man was extracted by the Dade County (FL) and Fairfax County (VA) Fire/Rescue USAR Task Forces following 13 days of entrapment in a collapsed hotel:

he had a broken ankle and was dehydrated. A man, Buck Helm, was pulled from the collapsed Nimitz Freeway by California rescuers four days after the 1989 Loma Prieta earthquake. (He subsequently died in the hospital from complications of crush syndrome, a common entrapment-related malady that may be effectively treated by paramedics and physicians during and after the extrication process—an important lesson for fire/rescue and medical professionals faced with long-term entrapment of victims in collapsed buildings.)

This knowledge, combined with experience gleaned from other earthquakes and collapse incidents, accounted for the decision to conduct round-the-clock search and rescue operations for nearly 16 days following the 1995 bombing of the Alfred P. Murrah Building in Oklahoma City and will have a profound effect on long-term search and rescue operations following major earthquakes in the coming century. . . .

The level of experience needed to properly evaluate rescue opportunities and conduct effective search and rescue in collapse disasters is a rare commodity. Fire department collapse rescue training courses, as well as classes taught to FEMA [Federal Emergency Management Agency] USAR Task Forces, provide the basis for this expertise. The United States, where seminal earthquake response training originated during the 1970s and the 1980s, is host to one of the most comprehensive systems of earthquake-related emergency training in the world. But as good as this training is for laying the groundwork for the nation's firefighters and rescuers, nothing beats hands-on experience conducting emergency operations in actual collapse disasters ranging from earthquakes to explosions.

Because collapse disasters are relatively rare, most firefighters gain hands-on experience only if they live or work in disaster-prone areas. Notable exceptions are FEMA USAR Task Forces, which may be dispatched to disasters anywhere in the United States; USAR teams sponsored by the State Department's Office of Foreign Disaster Assistance (part of the Agency for International Development); and similar teams sponsored by nations such as Israel, Switzerland, and Japan.

Consequently, teams of highly experienced rescuers are often scarce at earthquake disasters around the world, and locals are often left to their own devices, which may not include tools and methods designed to locate victims, stabilize structures, do tunneling, and lift heavy materials from collapsed buildings. In many cases, these countries' emergency responders lack the experience to recognize that many people may be trapped alive within piles of rubble and inside badly damaged buildings. As a result, untold numbers of trapped victims have been abandoned for dead when officials prematurely declared an end to the search and rescue phase and the beginning of the so-called recovery phase—often shorthand for "we are going to begin bulldozing buildings with heavy equipment with little regard for the potential of survivors trapped within them." This mindset should pose concern for residents of quakeprone regions, especially if local emergency officials share the same belief. Human survival times of one to two weeks (and sometimes beyond) should be the expected result of damaging quakes. . . .

Planning for Sustained Search and Rescue Operations

In the wake of the August 1999 Turkey earthquake, it was reported that some search and rescue teams began packing up to leave after just a few days. Many experienced fire and rescue professionals were puzzled by these reports, especially considering the tremendously high number of people who were missing and presumed trapped in collapsed structures across Turkey. Here is reinforcement of a lesson learned long ago by many fire and rescue professionals from Southern California: Emergency officials should be planning to pull live victims from the rubble for up to three weeks after catastrophic earthquakes, and they should be prepared to sustain nonstop search and rescue operations until all hope of locating viable victims has passed. The public has the right to expect this level of response to devastating quakes in the United States.

Once the victims who are obviously visible, lightly trapped, and readily reachable are rescued, resist the temptation to declare the start of the recovery phase until potential survivable void spaces have been searched for live victims. This is accomplished through two standard USAR processes known as void space search and selective debris removal.

During the void space search phase, well-trained and equipped firefighters and USAR task forces tunnel their way through the building using special tools, rope rescue, mining and tunneling, and structural stabilization methods. They use fiber-optic and ground-penetrating radar technology, special search cameras, extremely sensitive acoustic- and vibration-sensing instruments, search dogs, and direct visual and voice contact to locate victims trapped within void spaces created when the structure collapsed. In many cases, firefighters and other rescuers must squeeze through cracks and void spaces and crawl through the interior of collapsed buildings to positively determine whether victims are trapped. This work is extremely hazardous, but it is essential. This was the method used to locate many victims within the collapsed Northridge Meadows apartments on January 17, 1994.

After all known survivable void spaces are searched, selective debris removal begins. Firefighters and USAR task forces work closely with heavy equipment operators, structural engineers, construction and demolition contractors, and others to take the building apart piece by piece, usually from top to bottom. As the building's upper layers are selectively peeled away like an onion, newly accessible parts of the building are checked for potential survivors.

Alternating void space searches with selective debris removal should generally continue until all void spaces have been checked, the entire building has been dismantled, or all possible survivors have been located and extracted. These operations are extremely dangerous because of the instability of damaged buildings, as well as the continuing aftershocks that accompany major earthquakes. Without proper training, equipment, and experience, personnel conducting these operations can cause the building to collapse, killing rescuers and victims alike.

During these stages of a disaster, some of the most difficult, complex, and time-consuming rescues are made. These operations are the "bread and butter" operations of modern USAR-ready fire departments, as well as FEMA's 27 USAR Task Forces. They are also a mainstay of many internationally deployed USAR teams. Until these phases of rescue have been completed, officials should refrain from declaring that the "recovery phase" has begun.

In the Taiwan disaster, two pet dogs were found alive in the rubble of their collapsed apartment building 18 days after the earthquake struck. Apparently, the dogs were trapped beneath furniture that created a survivable void space and prevented the ceiling from completely flattening the apartment. They survived by drinking water from an aquarium and eating thawing meat from a freezer ripped open by the collapse. If two dogs can survive 18 days trapped in a collapsed building, couldn't humans do the same under favorable conditions?

When dealing with serious trauma, the Golden Hour is often emphasized—the optimum survival time within which firefighters and paramedics attempt to deliver injured victims to the definitive care of trauma centers and hospitals, often by helicopter. In recent years, another critical trauma benchmark, sometimes referred to as the "Golden First Day" of structure collapse, has become prevalent—a reference to the fact that the survival rate of trapped people begins to drop off after the first day. But, given what is known today, no one should assume that the situation is a lost cause after the first day. Victims missing within collapsed buildings should not be written off simply because 24 hours have slipped by. Based on the current data, it is clear that search and rescue operations should be scaled back only after all potential survivable void spaces have been inspected, even if it takes two or three weeks.

Disastrous earthquakes around the world have been notorious for disorganized emergency response, communications, a lack of appropriate tools, inexperienced rescuers, stifled decision making, and the misinformed or misdirected judgment of some officials who took it upon themselves to terminate search and rescue operations prematurely to start "recovery operations" while live victims await assistance in vain within the rubble of collapsed buildings.

In some cases search and rescue efforts were called off when it became embarrassingly obvious that the local jurisdiction or country had to rely on outside search and rescue assistance, and when certain political leaders—or their constituents—viewed the acceptance of outside help as a sign of weakness in their own systems.

Of course, it is the responsibility of U.S. fire and rescue professionals and public officials to ensure that such mistakes are not repeated. Fortunately, many earthquake-prone regions are blessed with public officials who support fire and rescue efforts during times of disaster, and who do not consider using outside assistance as a sign of weakness. They realize that mutual aid makes sense to ensure the most effective life- and property-saving service during widespread emergencies and that disaster search and rescue operations may be lengthy.

California's Earthquake Preparedness Efforts

By Janet Ward

Earthquake-prone California is active in earthquake preparedness and management, as Janet Ward, editor of American City & County, *explains in the following selection. Following the 1989 quake which devastated the California city of Loma Prieta, the state legislature passed an act that required state geologists not only to map seismic hazard zones but to evaluate these hazard areas before construction could proceed in them. A consortium of California cities, towns, and counties combined resources and suggested plans to help local governments minimize the disastrous effects of earthquakes. The consortium used a highly accurate geographic information system (GIS) to help produce better seismic hazard maps. Seismologists say these maps are the connections between earthquake research and the minimization of earthquake hazards. The United States Geological Survey has been publishing earthquake hazard maps since 1976 as an earthquake preparedness aid for city planners, engineers, emergency services, and the general public.*

Residents of San Francisco had been expecting The Big One for years. And while they all agreed that Loma Prieta, which devastated the city in 1989, was not it, they certainly had to admire its timing.

The magnitude 7.1 earthquake struck as the San Francisco Giants and the Oakland A's were preparing to take the field in what

Janet Ward, "Mapping the Big One: GIS/GPS Helps Pinpoint Seismic Trouble Spots," *American City & County*, vol. 112, March 1997, p. 45. Copyright © 1997 by Primedia: Business Magazine and Media. Reproduced by permission.

would be the first game of the World Series. Television cameras from around the world, there to record something as non-threatening as a baseball game, were pressed into service to film the near-destruction of one of America's most beautiful cities.

Americans were riveted to TV sets showing buckled freeways, crushed buildings and fires raging through the city's tony Marina District.

But if geologists and seismologists working in California have their way, there will not be a rerun.

Technology Takes On Nature

Loma Prieta was a wakeup call. The disastrous losses galvanized the state legislature, which in 1990 passed the California Seismic Hazards Mapping Act.

That act required the state geologist to map seismic hazard zones in order to identify areas prone to ground failure. But, more importantly, the act requires evaluation of those hazards before construction projects in potentially affected areas may proceed.

Predicting earthquakes is a dicey proposition at best. Especially in California, where fault lines criss-cross like a diagram of the New York subway system, the best that seismologists can offer is, "Stay here long enough, and you will experience an earthquake."

Still, you can't prepare if you can't predict, and preparation is what it's all about.

That, at least, is the philosophy behind research being done by the Association of Bay Area Governments (ABAG), a consortium of cities, towns and counties that combine resources to address issues of local concern.

With funding from the National Science Foundation and the Interior Department's U.S. Geological Survey (USGS), ABAG has been using Geographic Information System (GIS) to produce hazard maps that pinpoint potential trouble spots.

Jeanne Perkins, ABAG's Earthquake Program Manager, has spearheaded this effort since 1975. With the data provided by these maps, she and Jack Boatwright, a seismologist with the USGS, produced a comprehensive report called "On Shaky Ground," that was designed to be the definitive work on earthquake preparedness in the Bay Area. Perkins has produced a virtual library of reports for ABAG on everything from local government liability to seismic retrofits.

Additionally, ABAG offers mitigation options that can help local governments minimize the effects of an earthquake on their populations. Such mitigation options include:

- land use and zoning controls, particularly for critical or hazardous facilities;
- special building design requirements;
- disaster response planning;
- infrastructure and lifeline requirements; and
- programs to strengthen housing.

ABAG's files include—or soon will include—maps that designate fault traces, fault study zones, ground-shaking intensity, dam failure inundation areas and tsunami inundation areas.

Additionally, the U.S. Census Bureau's TIGER files, which include streets, city boundaries, county boundaries and census tract boundaries, can be combined with ABAG's files.

Hazard Management

GIS and Global Positioning System (GPS) have become an indispensable aspect of what experts call "hazard management," so indispensable in fact that they are often linked in engineering courses on the subject. The Hazards Research Laboratory at the University of South Carolina in Columbia, for instance, bills itself as "one of the primary centers of GIS/hazards research in the nation" and maintains a directory of researchers using or investigating the use of GIS in hazards/disaster management. Likewise, the National Information Service for Earthquake Engineering at the University of California-Berkeley has established a discussion list for anyone interested in GIS use in hazards management and research, especially that involving earthquakes.

"In California," says Michael Scott, a PhD candidate in geography at South Carolina, "GIS consultants and providers are making earthquakes their business."

One of those is Paul Wilson, president of Dallas-based MapFrame, which has provided GIS mapping and analysis software to ABAG for years. Wilson was ABAG's principal planner and chief of its technical information division when the group began exploring the possibilities of mapping earthquakes with GIS. He and Perkins began building the database that would become ABAG's technological foundation, but Proposition 13, California's infamous tax initiative, resulted in funding cuts that left the association unable to sustain their work.

After unsuccessfully shopping the data to other agencies, Wilson decided that the only feasible way to maintain it was to start his own company.

Now MapFrame markets GIS software to a number of local governments, and Wilson, an outspoken admirer of Perkins, continues to work with ABAG because "it's work that is done for a good reason."

"Seismic hazard maps are the connection between earthquake research and the mitigation of earthquake hazards," says Art Frankel, a seismologist with USGS and director of the National Seismic Hazard Mapping Project, a joint project of the USGS and the California Division of Mines and Geology. "These maps convey the ground motions that have a specified chance of occurring over a certain period of time."

The USGS has been publishing "probabilistic" maps for earthquake hazards since 1976. These maps have served as the basis for seismic zonation in model building codes and are used by city and county planners, engineers, emergency services departments and the general public.

"Our goal is damage minimization," Frankel says. "A major point of making these maps is to provide information on expected ground shaking so that buildings, bridges and other structures can be designed to withstand any expected shaking."

GIS has been particularly useful, according to Frankel, in creating hybrid maps that combine maps of expected ground motion in specific areas with map layers depicting population density, building type and other factors (dams, utility lines, the potential for tsunamis) that would have direct bearing on the potential for catastrophe.

Of course, the accuracy that GIS brings to the process is a big plus as well. ABAG's mapping system is based on grid cells 100 meters by 100 meters—2.1 million of them in the Bay Area. The resulting maps can contain a hundred different layers of information, different combinations based on what seismologists think is more likely to happen.

"They call that the 'largest credible earthquake' on that particular fault," says MapFrame's Wilson. "You layer it with maps that include all the bad stuff that can get you."

Additionally, the maps are easily convertible to other useful formats. For instance, Perkins handled a project for the California Department of Transportation (Caltrans) in which she produced a file of bridges and overpasses that was then overlaid on an earthquake model. Caltrans could then determine the vulnerability of various overpasses and bridges in various earthquake scenarios and plan its retrofitting programs.

BARD

But ABAG and the USGS are not the only initials involved in helping plan mitigation programs for the Bay Area. The Bay Area Regional Deformation Network (BARD) is keeping close track of strain accumulation along Bay Area faults, and again GIS and GPS are playing a major role. In fact, GPS receivers from Ashtech and Trimble Navigation, two Sunnyvale, Calif., providers, form the basis of the BARD system, which covers a 200-mile area stretching from the Sierra Nevada foothills to the Farallon Islands. BARD is a collaborative effort of UC-Berkeley's seismographic station, the USGS, Stanford University, the University of California-Davis, the University of California-Santa Cruz, Trimble, Ashtech, the Lawrence Livermore National Laboratory, the U.S. Coast Guard, the National Geodetic Survey, the Scripps Institute of Oceanography and the Jet Propulsion Lab. Its purpose is mapping crustal deformation (movements in the earth's crust that can be earthquake predictive). GPS stations, mounted in rock for stability, receive information from orbiting satellites and record that data every 30 seconds, 24 hours a day. Minute changes—"equal to the rate at which a fingernail grows," according to Seismographic Station Chief Engineer Bill Karavas—are recorded. This represents a giant leap forward for the seismologists, who formerly relied on analog 'strain' or 'creep' meters, which measured changes over a 100-foot stretch of land that were then extrapolated for a larger area.

"Ultimately," Karavas says, "we hope to be able to use all this data to predict earthquakes."

"Previously there was no way to get this type of data over wide areas," says Ashtech's Mark Eustis. "Six or seven years ago, you had to buy a mainframe computer. Now, using the web, you can go to the Economic and Social Research Institute's (ESRI's) home page, pull down maps and put them into a desktop PC that runs complete, usable GIS software. It's brought affordability to the local researcher level."

It has also made Californians safer and ensured local governments "safer communities [now] and stronger economies following future earthquakes," as Perkins says.

The Use of the Internet in Earthquake Response Efforts

BY DAVID WALD, LISA WALD, JAMES DEWEY,
VINCE QUITORIANO, AND ELISABETH ADAMS

*The United States Geological Survey (USGS) has a website for the
creation of "Community Internet Intensity Maps." What used to take
months through mailed questionnaires can now be done in a matter of
hours using the Internet. Anyone experiencing an earthquake can go to
the website and answer questions pertaining to the intensity and magnitude of what they experienced. From their responses, experts draw up a
color-coded map showing where and how strongly the earthquake was
felt, allowing them to quickly assess its severity and the potential for aftershocks. This system may be the only means available to gather such data
in areas where there are few seismic instruments, and the system can assist
all areas with more effective emergency response and preparations for future earthquakes. David Wald is a seismologist for the USGS. Lisa Wald
is a geophysicist and outreach and education director for the USGS.
James Dewey is a scientist for the USGS National Earthquake Information Center. Vince Quitoriano works with the USGS Western Region
Western Earthquake Hazards Team. Elisabeth Adams is a geophysics
student and was a programmer and researcher for "Did You Feel It?"*

Since the early 1990's, the magnitude and location of an
earthquake have been available within minutes on the Internet. Now, as a result of work by the U.S. Geological Survey (USGS) and with the cooperation of various regional seis-

David Wald, Lisa Wald, James Dewey, Vince Quitoriano, and Elisabeth Adams,
"Did You Feel It? Community-Made Earthquake Shaking Maps," *USGS Fact Sheet
030-01*, 2001.

mic networks, people who experience an earthquake can go online and share information about its effects to help create a map
of shaking intensities and damage. Such "Community Internet
Intensity Maps" (CIIM's) contribute greatly in quickly assessing
the scope of an earthquake emergency, even in areas lacking seismic instruments.

Not long ago, the first thing that most people did after feeling
an earthquake was to turn on their radio for information. Recently, however, after the Hector Mine earthquake in southern
California and after a widely felt earthquake in upstate New
York, many people logged onto the Internet instead, not only to
get information, but also to share their own experience of the
earthquake. After checking the U.S. Geological Survey (USGS)
website for the location and magnitude of the earthquake, they
went to a webpage called "Did You Feel It?" (http://pasadena.wr.
usgs.gov/shake/). They entered their ZIP code and answered a
list of questions such as "Did the earthquake wake you up?" and
"Did objects fall off shelves?" In minutes a map began taking
shape on the Internet, and in a couple of hours, with more than
several thousand responses for the southern California event, a
Community Internet Intensity Map (CIIM) showed where and
how strongly the earthquake had been felt.

Modified Mercalli Intensities

The measure of shaking and damage from an earthquake is called
"seismic intensity." In general, the intensity decreases with increasing distance from the earthquake. However, a variety of factors, such as the type of earthquake, rupture direction, local geography, soil conditions, and type and age of buildings, result in
an often complicated pattern of varying intensities from place to
place. Since 1931, the USGS has been assigning intensities to
earthquakes in the United States on the basis of the Modified
Mercalli Intensity (MMI) scale. Typically, this is done by collect-
ing responses to a postal questionnaire that is sent to each post
office near the earthquake, and to a sparser sample of post offices
with increasing distance from the earthquake. This way of
preparing a seismic intensity map can take months to complete.

Community Internet Intensity Maps

In contrast to the old method, CIIM's take advantage of the Internet to generate initial intensity maps almost instantly. Data are

received through a questionnaire on the Internet answered by people who actually experienced the earthquake, reducing the process of preparing and distributing a shaking intensity map from months to minutes.

A CIIM summarizes the responses, and an intensity number is assigned to each ZIP code for which a CIIM questionnaire is completed. The intensity values in each ZIP-code area are averaged, and the map is updated as additional data are received. ZIP-code areas for which data have been received are color-coded according to the intensity scale below the map; other areas are gray.

A CIIM is automatically made after each widely felt earthquake in the United States. The system can start receiving responses within about 3 minutes after the earthquake. Internet users can also enter data for U.S. earthquakes they have experienced in the past.

CIIM values have been calibrated to be, on average, similar to MMI values. Comparisons between traditional MMI values obtained by using postal questionnaires and CIIM values show that they agree well except in areas of low shaking, where the CIIM values tend to be more reliable.

"Did You Feel It?"

In regions with few earthquakes and, hence, few seismic instruments, which includes most of the United States and most of the world, large numbers of intensity observations for a small to moderate event can indicate which areas will be more prone to shaking in the less frequent larger earthquakes. After a damaging earthquake in those sparsely instrumented areas, CIIM's can provide information about which areas experienced the most shaking and therefore the most potential damage. This information can serve as a post-earthquake response tool and for estimating losses from future earthquakes.

In areas such as California where there are networks of seismic instruments, CIIM's provide a very rapid means of displaying the pattern of shaking independently of strong-motion seismographs. CIIM's provide descriptions of actual damage, rather than inferred potential damage indicated by instrumental shaking records. Also, the potential number of Internet responses far exceeds the number of seismic instruments, so very dense sampling of earthquake effects can be collected, providing details that would not be possible with the instruments alone. Eventually, the automated use of voluntary Internet contributions for the collection of intensity data following an earthquake will significantly reduce the manpower required to collect and interpret intensity observations.

Finally, the interactive nature of the Internet-based questionnaire and mapping provides an unprecedented opportunity for community involvement. The CIIM interactive website provides an avenue for feedback among the communities affected by earthquakes, the scientists studying their effects, and agencies responding to the events. The CIIM website allows people in an area struck by an earthquake to share their experiences, and this type of communication after a stressful event is an important tool for dealing with the emotional impact of such an experience.

Most areas of the United States, even many of those prone to earthquakes, do not have dense seismic networks. For such areas, CIIM's provide the only rapid way to assess the distribution of shaking intensities and levels of damage. This quick information can aid in making the most effective use of emergency response resources and assist in preparing for future earthquakes. The work of the USGS and its cooperators in developing CIIM's are only part of the ongoing USGS efforts to protect lives and property from future earthquakes throughout the United States.

The Most Destructive Earthquakes on Record

Date	Location	Deaths
January 23, 1556	China, Shansi	830,000
July 27, 1976	China, Tangshan	255,000, later estimates 650,000–750,000
August 9, 1138	Syria, Aleppo	230,000
May 22, 1927	China, near Xining	200,000
December 22, 856	Iran, Damghan	200,000
December 16, 1920	China, Gansu	200,000
March 23, 893	Iran, Ardabil	150,000
September 1, 1923	Japan, Kwanto	143,000
December 28, 1908	Italy, Messina	70,000 to 100,000
September 1290	China, Chihli	100,000
November 1667	Caucasia, Shemakha	80,000
November 18, 1727	Iran, Tabriz	77,000
November 1, 1755	Portugal, Lisbon	70,000
December 25, 1932	China, Gansu	70,000
May 31, 1970	Peru	66,000
1268	Asia Minor, Silicia	60,000
January 11, 1693	Italy, Sicily	60,000
May 30, 1935	Pakistan, Quetta	30,000 to 60,000
February 4, 1783	Italy, Calabria	50,000
June 20, 1990	Iran	50,000

The Largest Earthquakes in the United States

Magnitude	Date	Location
9.2	March 28, 1964	Prince William Sound, Alaska
8.8	March 9, 1957	Andreanof Islands, Alaska
8.7	February 4, 1965	Rat Islands, Alaska
8.3	November 10, 1938	east of Shumagin Islands, Alaska
8.3	July 10, 1958	Lituya Bay, Alaska
8.2	September 10, 1899	Yakutat Bay, Alaska
8.2	September 4, 1899	near Cape Yakataga, Alaska
8.0	May 7, 1986	Andreanof Islands, Alaska
7.9	February 7, 1812	New Madrid, Missouri
7.9	January 9, 1857	Fort Tejon, California
7.9	April 3, 1868	Ka'u District, Island of Hawaii
7.9	October 9, 1900	Kodiak Island, Alaska
7.9	November 30, 1987	Gulf of Alaska

The Largest Earthquakes in the Contiguous United States

Magnitude	Date	Location
7.9	February 7, 1812	New Madrid, Missouri
7.9	January 9, 1857	Fort Tejon, California
7.8	March 26, 1872	Owens Valley, California
7.8	February 24, 1892	Imperial Valley, California
7.7	December 16, 1811	New Madrid, Missouri area
7.7	April 18, 1906	San Francisco, California
7.7	October 3, 1915	Pleasant Valley, Nevada
7.6	January 23, 1812	New Madrid, Missouri
7.5	July 21, 1952	Kern County, California
7.3	November 4, 1927	west of Lompoc, California
7.3	December 16, 1954	Dixie Valley, Nevada
7.3	August 18, 1959	Hebgen Lake, Montana
7.3	October 28, 1983	Borah Peak, Idaho

active fault: A fault along which a slip has occurred or earthquake foci are located.

aftershocks: Smaller earthquakes following the largest earthquake of a series concentrated in a restricted crustal volume.

amplitude: The maximum height of a wave crest or depth of a trough.

aseismic region: A region almost free of earthquakes.

asthenosphere: The soft layer below the lithosphere that is probably partially molten.

body wave: A seismic wave that travels through the interior of an elastic material; P and S seismic waves.

continental shelf: The offshore area of a continent in a shallow sea.

core: The central part of the earth below a depth of 2,900 kilometers.

crust: The outermost rocky shell of the earth.

density: The mass per unit volume of a substance.

dip: The angle by which a rock layer or fault plane deviates from the horizontal.

earthquake: The vibrations of the earth caused by the passage of seismic waves radiating from some source of elastic energy.

elastic rebound theory: The theory of earthquake generation proposing that faults remain locked while strain energy slowly accumulates in the surrounding rock and then suddenly slip, releasing this energy in the form of heat and seismic waves.

epicenter: The point on the earth's surface directly above the focus (or hypocenter) of an earthquake.

fault: A fracture or zone of fractures in rock along which the two sides have been displaced relative to each other and parallel to the fracture.

fault plane: The plane that most closely coincides with the rupture surface of a fault.

focal depth: The depth of the earthquake's focus below the surface of the earth.

focus: The place at which rupture commences.

foreshocks: Smaller earthquakes preceding the largest earthquake of a series concentrated in a restricted crustal volume.

frequency: The number of oscillations per unit of time where the unit is hertz (Hz), which equals 1 cycle per second.

inner core: The central solid region of the earth's core.

intensity: A measure of ground shaking obtained from the damage done to structures built by humans, changes in the earth's surface, and public reports.

interplate earthquake: An earthquake with its focus on a plate boundary.

intraplate earthquake: An earthquake with its focus within a plate.

island arc: A chain of islands above a subduction zone (e.g., Japan, Aleutians).

liquefaction: The process in which soil behaves like a dense fluid rather than a wet solid mass during an earthquake.

lithosphere: The outer, rigid shell of the earth above the asthenosphere that contains the crust, continents, and plates.

Love waves: Seismic surface waves with only horizontal shear motion transverse to the direction of propagation.

magnitude: A measure of earthquake size determined by taking the common logarithm (base 10) of the largest ground motion recorded during the arrival of a seismic wave type and applying a standard correction for distance to the epicenter. Three common types of magnitude are Richter (or local), moment, and surface wave.

mantle: The main bulk of the earth between the crust and core, ranging from depths of about 40 to 3,470 kilometers.

moment magnitude: The magnitude of an earthquake estimated by using the seismic moment.

outer core: The outer liquid shell of the earth's core.

plate: A large, relatively rigid segment of the earth's lithosphere that moves in relation to other plates over the deeper interior. Plates meet in convergence zones and separate at divergence zones.

plate tectonics: A geological model in which the earth's crust and uppermost mantle (lithosphere) are divided into a number of more-or-less rigid segments (plates).

prediction: The forecasting in time, place, and magnitude of an earthquake or of strong ground motions.

P wave: The primary or fastest wave traveling away from a seismic event through the rock, consisting of a train of compressions and expansions of the material.

Rayleigh waves: Seismic surface waves with ground motion only in a vertical plane.

rift: A region where the crust has split, usually marked by a rift valley (e.g., East African Rift, Rhine Graben).

risk: The probability of loss of life and property from an earthquake hazard within a given time interval and region.

scarp: A cliff or steep slope formed by displacement of the ground surface.

seismic gap: An area in an earthquake-prone region where there is a below-average release of seismic energy.

seismicity: The occurrence of earthquakes in space and time.

seismic wave: An elastic wave in the earth, usually generated by an earthquake source or an explosion.

seismograph: An instrument for recording as a function of time the motions of the earth's surface that are caused by seismic waves.

seismology: The study of earthquakes, seismic sources, and wave propagation through the earth.

slip: The relative motion of one face of a fault relative to the other.

subduction zone: A region where an oceanic plate dives below a continental plate into the mantle; ocean trenches are the surface expression of a subduction zone.

surface wave: A seismic wave that follows the earth's surface with a speed less than that of S waves. There are two types of surface waves: Rayleigh waves and Love waves.

surface-wave magnitude: The magnitude of an earthquake estimated from measurements of the amplitude of surface waves.

swarm: A series of earthquakes in the same locality with no earthquake of outstanding size.

S wave: The secondary seismic wave traveling more slowly than the P wave, consisting of elastic vibrations transverse to the direction of travel.

tectonic earthquake: An earthquake resulting from the sudden release of energy stored by a major deformation of the earth.

tectonics: The history of the earth's larger features (rock formations and plates) and the forces and movements that produce them.

tomographic: The construction of the image of an internal object or structure from measurements of seismic waves at the surface.

trench: A depression on the ocean floor caused by plate subduction.

tsunami: A long ocean wave usually caused by seafloor displacement in an earthquake.

FOR FURTHER RESEARCH

Books

Bruce A. Bolt, *Earthquakes and Geological Discovery*. New York: Freeman, 1993.

C. Davison, *The Founders of Seismology*. Cambridge, England: Cambridge University Press, 1927.

Earthquake Preparedness Society, *Earthquakes and Preparedness: Before, During, and After*. Downey, CA: Earthquake Preparedness Society, 1990.

James M. Gere and Haresh C. Shah, *Terra Non Firma*. Stanford, CA: Stanford Alumni Association, 1991.

B. Gutenberg and Charles F. Richter, *Seismicity of the Earth and Associated Phenomena*. NJ: Princeton University Press, 1954.

Robert M. Hamilton and Arch C. Johnston, *Tecumseh's Prophecy: Preparing for the Next New Madrid Earthquake*. U.S. Geological Survey Circular 1066, 1986.

Eugene J. Hass and Denise S. Mileti, *Socioeconomic Impact of Earthquake Prediction on Government, Business, and Community*. Boulder: Institute of Behavioral Sciences, University of Colorado, 1976.

Thomas A. Heppenheimer, *The Coming Quake: Science and Trembling on the California Earthquake Frontier*. New York: Times Books, 1988.

Philip Kearey and Frederick J. Vine, *Global Tectonics*. Oxford, England: Blackwell Scientific Publications, 1990.

Edward A. Keller and Nicholas Pinter, *Active Tectonics, Earthquakes, Uplift, and Landscape*. Englewood Cliffs, NJ: Prentice-Hall, 1996.

Virginia Kimball, ed., *Earthquake Ready: The Complete Preparedness Guide,* expanded and updated. Malibu, CA: Roundtable, 1992.

Pierre Kohler, *Volcanoes and Earthquakes.* New York: Barron, 1987.

Matthys Levy and Mario Salvadori, *Why the Earth Quakes.* New York: W.W. Norton, 1995.

Ron L. Morton, *Music of the Earth: Volcanoes, Earthquakes, and Other Geological Wonders.* New York: Plenum Press, 1996.

Eric J. Oliver, *Shocks and Rocks: Seismology and the Plate Tectonics Revolution.* Washington, DC: American Geophysical Union, 1996.

Frank Press and Raymond Siever, *Earth.* New York: Freeman, 1998.

Gregory Vogt, *Predicting Earthquakes.* New York: Watts, 1989.

Bryce Walker, *Earthquake.* Alexandria, VA: Time–Life Books, 1982.

Robert M. Wood, *Earthquakes and Volcanoes.* New York: Grove Press, 1987.

Peter Yanev, *Peace of Mind in Earthquake Country.* San Francisco: Chronicle Books, 1991.

Periodicals

C.J. Anderson, "Animals, Earthquakes, and Eruptions," *Field Museum of Natural History Bulletin*, 1973.

Madonna Aveni, "Computer Network Provides Real-Time Earthquake Data," *Civil Engineering*, June 2001.

Lisa Beyer and Andrew Finkel, "Seeking Survival and More," *Time*, September 1999.

Harvey Black, "Underworld Connections," *Earth*, August 1998.

Bruce A. Bolt, "Balance of Risks and Benefits in Preparation for Earthquakes," *Science*, January 1991.

Pedro D. Botta, "The Day the Earth Shook," *Hemisphere: A Magazine of the Americas*, Fall 1997.

Keay Davidson, "Superdeep Earthquakes," *Earth*, November 1994.

Joshua Fischman, "Falling into the Gap," *Discover*, October 1992.

Cliff Frohlich, "Deep Earthquakes," *Scientific American*, January 1989.

Arch C. Johnston, "A Major Earthquake Zone on the Mississippi," *Scientific American*, April 1982.

Beth Kephart, "The Rubble-Rouser," *Americas*, May/June 2001.

A.G. Lindh, "Earthquake Prediction Comes of Age," *Technology Review*, February/March 1990.

Steve Nadis, "Detecting Earthquakes from Space," *Technology Review*, November/December 1997.

Susan Oh and Laurie Udesky, "Terror in Turkey," *Maclean's*, August 1999.

Carol S. Prentice and David K. Keefer, "Surface Effects of the Earthquakes," *Earthquakes and Volcanoes*, 1992.

Tim Radford, "Lessons from Past Disasters Go Unheeded," *Geodate*, March 2001.

Mark B. Roman, "Finding Fault," *Discover*, August 1988.

Linda Rowan, "Where the Ground Shakes," *Science*, March 2000.

Ross S. Stein and Robert S. Yeats, "Hidden Earthquakes," *Scientific American*, June 1989.

Shawna Vogel, "Shocks Heard Round the World," *Discover*, January 1990.

Hugh Westrup, "Black Saturday," *Current Science*, March 2001.

Websites

Earthquake Hazards Program, http://earthquake.usgs.gov. This website has information on worldwide earthquake activity, earthquake science, and earthquake hazard reduction.

EQNet, www.eqnet.org. The Earthquake Information Network provides current earthquake information and regional and state information.

Federal Emergency Management Agency, www.fema.gov. The FEMA homepage has links to other FEMA publications regarding earthquake engineering and seismic safety.

IRIS Seismic Monitor, www.iris.edu. This is an interactive display of global seismic activity where users can monitor earthquakes and visit worldwide seismic stations.

National Earthquake Information Center, www.neic.usgs.gov. The NEIC of the U.S. Geological Survey, located in Golden, Colorado, is a gateway for worldwide earthquake information.

Seismological Society of America, www.seismosoc.org. This is an international association devoted to the advancement of earthquake science; the site provides many links.

Understanding Earthquakes, www.crustal.ucsb.edu. Understanding Earthquakes is a student-friendly site featuring a rotating globe showing earthquake locations, famous earthquake accounts, the history of seismology, and links to other educational earthquake sites.

United States Geological Survey, http://quake.usgs.gov. The USGS site presents information on earthquake hazards and preparedness.

University of Alaska Geophysical Institute, www.giseis.alaska.edu. This site provides information on Alaskan earthquakes, active faults, recent earthquakes, and earthquake press releases.

University of California at Berkeley Seismological Laboratory, www.seismo.berkeley.edu. This site has recent earthquake information and instructions on how to make your own seismogram.

INDEX